George F. Hampson

On the Classification of the Chrysauginae, a Subfamily of Moths of the Family Pyralidae

George F. Hampson

On the Classification of the Chrysauginae, a Subfamily of Moths of the Family Pyralidae

ISBN/EAN: 9783741132308

Manufactured in Europe, USA, Canada, Australia, Japa

Cover: Foto ©berggeist007 / pixelio.de

Manufactured and distributed by brebook publishing software
(www.brebook.com)

George F. Hampson

On the Classification of the Chrysauginae, a Subfamily of Moths of the Family Pyralidae

Genus MOROVA.

Morova, Wlk. xxxii. 523 (1865).

Palpi porrect, thickly scaled and reaching to the frons, which has a rounded prominence; antennæ of male somewhat thickened; tibiæ smoothly scaled. Fore wing with the costa arched towards apex, the outer margin excurved at middle; veins 3, 4, 5 well separated at origin; 6 from below upper angle; 8, 9 stalked. Hind wing with the outer margin excurved at middle; vein 3 from close to lower angle of cell; 5 from middle of discocellulars; 6, 7 from upper angle.

Fig. 26.

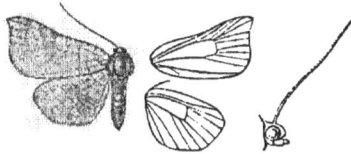

Morova subfasciata, ♂. ⅓.

Type. †MOROVA SUBFASCIATA, Wlk. xxxii. 523.   New Zealand; Fiji.
†*Cacoecia gallicolens*, Butl. Voy. Erebus & Terror, Ins. p. 46.

2. On the Classification of the *Chrysauginæ*, a Subfamily of Moths of the Family *Pyralidæ*. By Sir GEORGE F. HAMPSON, Bart., F.Z.S.

[Received April 8, 1897.]

The *Chrysauginæ* are a highly specialized subfamily of the true Pyralid group of the large family *Pyralidæ*, consisting in addition to the present subfamily of the *Epipaschianæ*, *Endotrichinæ*, and *Pyralinæ*, lately classified by me in the 'Transactions' of the Entomological Society, and characterized by vein 7 of the fore wing being stalked with 8, 9. The *Chrysauginæ* as here defined are primarily distinguished from their allies by the abortion of the maxillary palpi, which are well developed in almost all other *Pyralidæ*. They are closely allied to the *Endotrichinæ* but, as vein 8 of the hind wings is in rare instances free, were probably derived directly from the *Pyralinæ* as a parallel development to the *Endotrichinæ*. The latter are almost confined to the Old World, though a few species are found in the Nearctic region, and one genus in the W. Indies; whilst the *Chrysauginæ* are almost exclusively *Neotropical*, a few genera and species being found in the Southern States, and a few others spreading through the Australian region to the Malayan subregion, the furthest points reached being Burma and Assam.

The subfamily is remarkable for the great sexual diversity found in the subcostal neuration of the fore wing in a large proportion of the species, the females always having veins 7, 8, 9 stalked, as

is typical of the group. This diversity is usually correlated with
the development of various other secondary sexual characters, one
of the most common and remarkable being an ear-shaped tympanic
vesicle at the base of the costa of fore wing covered by a drum of
fine corrugated membrane.

A carefully elaborated classification of most of the genera was
published by the late E. M. Ragonot in the 'Annales de la Société
Entomologique de France'[1]; but a large number of the species
were unknown to him, and his material was insufficient for him to
discover the large amount of sexual dimorphism that exists. He
defined the *Chrysaugince* as differing from the *Endotrichince* in
being stoutly-built insects, and includes in the latter subfamily
many of the genera which by my definition fall into the former:
the paper, however, formed a most important contribution towards
a correct classification of the group, which was originally defined
and systematized by Lederer in 1863.

None of the genera are of a very generalized structure, but
Chrysauge itself, apart from its secondary sexual characters, is
regarded as the least specialized, with its short porrect palpi and
median nervules of both wings arising from the cell in its 1st section.
From it were developed a group of genera with downcurved palpi, of
which forms like *Anemosa* and *Pelasgis* have very long palpi; *Semnia*
with the palpi smooth and a tuft of hair on the antennæ; *Uliosoma*
and *Acutia* with one of the median nervules absent in one or both
wings; *Condylolomia* with veins 2 and 3 of the fore wing stalked;
Itambe and *Microzancla* with extremely falcate fore wings; *Macna*
with very long straight palpi in female, upturned and angled with
hair in front in male; *Psectrodes* and *Acrodegmia* with very long
palpi ending in a large rounded tuft of hair on 3rd joint.

Another large group of genera have the palpi upturned, of
which the majority have short palpi, such as *Sthenobœa* with vein 4
of hind wing absent; *Dasycnemia* with veins 4, 5 of both wings
stalked; *Anisothrix* without a frontal tuft; *Rucuma* with tufts of
hair on frons, mid tibiæ, and tarsal joints, and excisions in the
costa of fore wing; whilst a few genera have very long upturned
palpi, culminating in *Tamyra* with a rounded brush on 3rd joint.

A very curious structure found in several of the genera, of
which *Casuaria* is typical, is the development of the retinaculum
into a complete ring, the frenulum being thickened, flattened,
contorted at base and with a short lower fork; this form being
associated with a glandular swelling and tufts of hair on underside of costa of fore wing and the tympanic vesicle on upperside
mentioned above.

Subfamily CHRYSAUGINÆ.

Proboscis well-developed; palpi of extremely different forms in
the different genera; maxillary palpi absent; frons usually with a
tuft of hair. Fore wing with vein 7 stalked with 8, 9 in female;

---

[1] Ann. Soc. Ent. Fr. 1890, pp. 435-546; and 1891, pp. 15-114 & 559-662,
plates 5, 7, 8, & 16.

PHYLOGENY OF THE CHRYSAUGINÆ.

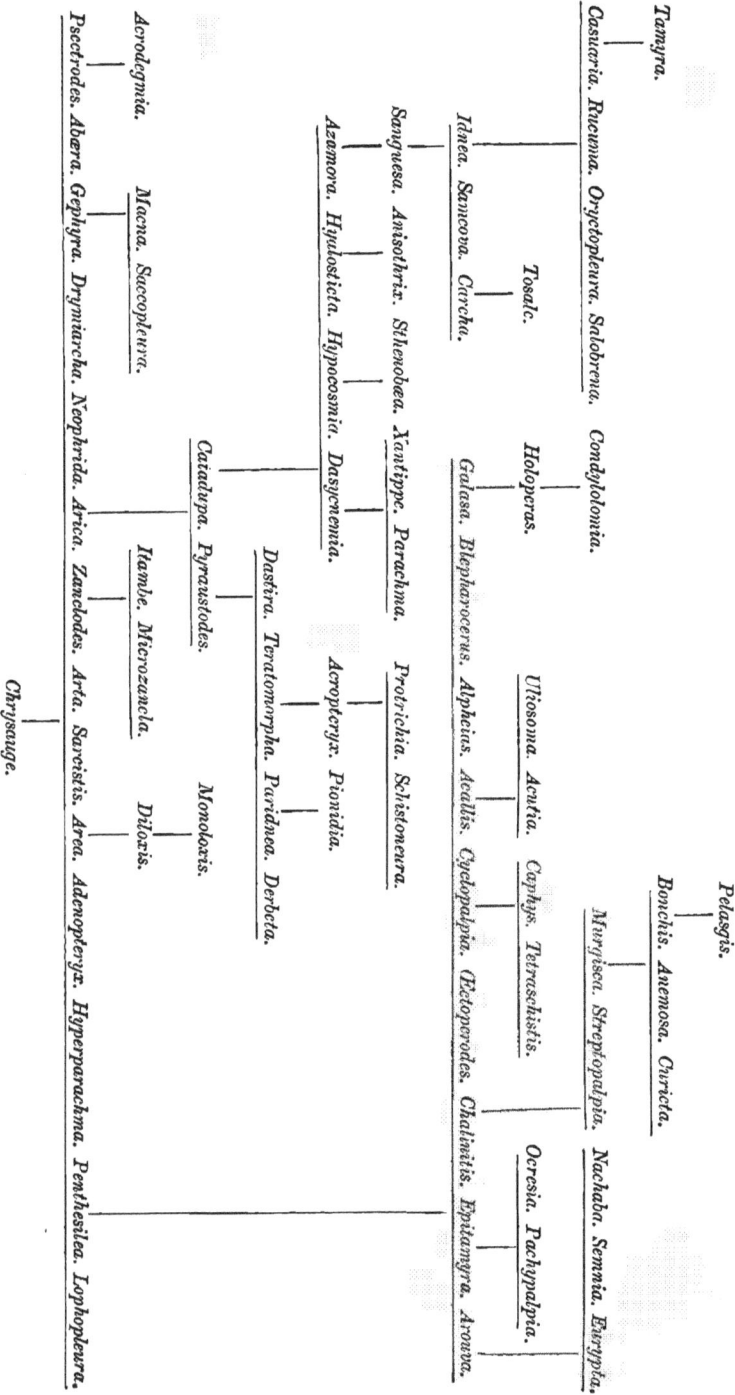

Tamyra.

Cascaria. Rucuma. Oryctopleura. Salobrena.

Itanca. Sancova. Corcha.

Tosalc.

Sanguesa. Anisothrix. Sthenobea. Xantippe. Parachma.

Azemora. Hyalosticta. Hypocosmia. Dasycnemia.

Aerodegmia.

Macna. Saccopleura.

Psectrodes. Abera. Gephyra. Drymiarcha. Neophrida. Arica. Zanclodes. Arta. Sarcistis. Area. Adenopteryx. Hyperparachma. Penthesilea. Lophopleura.

Chrysauge.

Condylolomia.

Holoperas.

Galasa. Blepharocerus. Alpheias. Acallis. Cyclopalpia. Œctoperodes. Chalinitis. Epitamyra. Aronva.

Ultosoma. Acutia.

Caphys. Tetraschistis.

Murgisca. Streptopalpia.

Oeresia. Pachypalpia.

Catacarpa. Pyraustodes.

Itamie. Microzancla.

Dastira. Teratomorpha. Paridnea. Derbeta.

Acropteryx. Pionidia.

Protrichia. Schistoneura.

Monoloxis.

Dioxis.

Pelagis.

Bonchis. Anemosa. Curicta.

Nachaba. Semnia. Energgia.

the subcostal neuration of male varying greatly in relation to the secondary sexual characters. Hind wing with the median nervure non-pectinate above; vein 7 almost always anastomosing with 8.

Key to the Genera.

A. Palpi upturned.
  a. Palpi about four times length of head, and reaching far above it.
    a^1. Palpi thickly scaled, the hair on 3rd joint forming a rounded brush.................... 9. *Tamyra*, p. 646.
    b^1. Palpi fringed with hair below throughout ...... 10. *Casuaria*, p. 647.
  b. Palpi about twice the length of head and reaching well above it.
    a^1. Palpi with the 2nd and 3rd joints strongly angled with hair in front......................... 4. *Macna* ($\male$), [p. 642.
    b^1. Palpi with the 3rd joint tufted with hair on inner side .... 13. *Salobrena*, p. 649.
    c^1. Palpi with the 3rd joint naked .... 15. *Samcova*, p. 651.
  c. Palpi with the 3rd joint not, or hardly, reaching above vertex of head.
    a^1. Fore and hind wings with veins 4, 5 from the cell.
      a^2. Palpi reaching vertex of head.
        a^3. Palpi with the 3rd joint fringed in front with long downcurved hair .... 63. *Curicta* ($\male$), [p. 682.
        b^3. Palpi with the 3rd joint not fringed with downcurved hair.
          a^4. Frons with a conical tuft of hair.
            a^5. Mid tibiæ and tarsal joints with large tufts of scales.
              a^6. Fore wing with vein 10 stalked with 7, 8, 9.
                a^7. Fore wing with vein 7 given off from 8 after 9 .... 16. *Tosale*, p. 652.
                b^7. Fore wing with vein 7 given off from 8 before 9.
                  a^9. Fore wing with the costa evenly curved.............................. 17. *Carcha*, p. 653.
                  b^9. Fore wing with the costa excised beyond middle .... 12. *Oryctopleura*, [p. 648.
              b^6. Fore wing with vein 10 free.
                a^7. Fore wing with the costa not excised beyond middle.
                  a^8. Hind wing with the discocellulars curved, the cell of moderate length .... 19. *Azamora*, p. 654.
                  b^8. Hind wing with the discocellulars strongly angled, the lower part of cell produced... 18. *Sanguesa*, p. 654.
              b^7. Fore wing with the costa excised beyond middle.
                a^8. Fore wing with the outer margin excised towards outer angle; male with glandular swelling on inner area of hind wing .... 11. *Rucuma*, p. 648.
                b^8. Fore wing with the outer margin not excised towards outer angle; male with glandular swelling at base of costa.. 14. *Idnea*, p. 650.

b^5. Mid tibiæ without large tufts of scales; the tarsal joints smooth.

a^6. Fore wing with the apex produced and acute; vein 11 free. [p. 663.

a^7. Hind wing with veins 6, 7 from cell. 31. *Acropteryx*,

b^7. Hind wing with veins 6, 7 stalked. 29. *Protrichia*, [p. 661.

b^6. Fore wing with the apex not produced; vein 11 anastomosing with 12. 30. *Schistoneura*, [p. 662.

b^4. Frons smooth.

a^5. Fore wing with the costa deeply excised beyond middle, the apex produced upwards . ... 32. *Teratomorpha*, [p. 663.

b^5. Fore wing with the costa evenly curved, the apex not produced.

a^6. Palpi closely applied to frons.

a^7. Palpi with the 3rd joint hidden in the hollow of a tuft at end of 2nd................. .... 20. *Anisothrix*, [p. 656.

b^7. Palpi with the 3rd joint exposed and thickly scaled; the tarsal joints with tufts of scales .... 21. *Hyalosticta*, [p. 656.

b^6. Palpi extending widely in front of frons, the 3rd joint naked.

a^7. Hind wing with vein 2 present... 34. *Paridnea*, p. 665.

b^7. Hind wing with vein 2 absent ... 33. *Pionidia*, p. 664.

b^2. Palpi reaching about halfway to vertex of head.

a^3. Frons with a conical tuft.

a^4. Fore wing long and narrow; antennæ of male pectinated............................. 35. *Derbeta*, p. 665.

b^4. Fore wing short and broad; antennæ of male ciliated .... 28. *Dastira*, p. 661.

b^3. Frons smooth .... 36. *Pyraustodes*, [p. 666.

b^1. Hind wing with vein 4 absent .... 22. *Sthenobæa*, [p. 657.

c^1. Hind wing with veins 4, 5 stalked.

a^2. Fore wing with veins 4, 5 from cell .... 23. *Hypocosmia*, [p. 657.

b^2. Fore wing with veins 4, 5 stalked.

a^3. Hind wing with vein 3 from angle of cell .. 26. *Dasycnemia*, [p. 660.

b^3. Hind wing with vein 3 absent .... 25. *Parachma*, [p. 659.

d^1. Hind wing with veins 3, 4, 5 stalked .... 24. *Xantippe*, p. 658.

B. Palpi with the 2nd joint upturned, the 3rd porrect; both wings with veins 4, 5 from cell.

a. Hind tarsi smooth; male with a tympanic vesicle. 37. *Arica*, p. 666.

b. Hind tarsi with a tuft of scales on 1st joint; male with no tympanic vesicle .... 27. *Catadupa*, [p. 660.

C. Palpi porrect.

a. Palpi straight and not downcurved at extremity.

a^1. Palpi hardly extending beyond the frons.

a^2. Hind wing with veins 4, 5 from cell.

a^3. Fore wing with vein 3 from cell.

a^4. Both wings with vein 2 present.

a^5. Fore wing with the costa not excised.. 76. *Chrysauge*, [p. 691.

b^5. Fore wing with the costa deeply excised beyond middle, the apex produced upwards... .... 40. *Zanclodes*, p. 668.

b^4. Both wings with vein 2 absent .... 47. *Hyperparachma*, [p. 672.

b^3. Fore wing with veins 3, 4, 5 stalked .... 75. *Lophopleura*,
b^2. Hind wing with veins 4, 5 stalked.   [p. 690.
   a^3. Hind wing with vein 3 present; palpi
     minute; the costa of fore wing excised and
     the apex produced upwards .... 38. *Itambe*, p. 667.
   b^3. Hind wing with vein 3 absent.   [p. 667.
     a^4. Fore wing with vein 3 from cell, the
       costa excised .... 39. *Microzancla*,
     b^4. Fore wing with veins 3, 4, 5 stalked,
       the costa not excised...................... 42. *Sarcistis*, p. 669.
  c^3. Hind wing with veins 3, 4, 5 stalked ...... 41. *Arta*, p. 669.
b^1. Palpi extending about twice the length of head.
  a^2. Palpi curved towards each other at tips;
    antennæ of male bipectinate, the basal joint
    with a hollow in front; fore wing with vein
    10 absent.. 7. *Drymiarcha*,
                                        [p. 645.
  b^2. Palpi straight; fore wing with vein 10 from 8. *Neophrida*,
    the cell ....   [p. 646.
  c^1. Palpi extending about three times length of
    head.
    a^2. Fore wing with the costa excised beyond
      middle; male with tympanic vesicle .... 6. *Gephyra*, p. 644.
    b^2. Fore wing with the costa evenly arched; male
      with no tympanic vesicle.
      a^3. Palpi with the 3rd joint fringed with
        hair below .... 3. *Abæra*, p. 641.
      b^3. Palpi with a rounded brush of hair on
        3rd joint; hind wing with vein 8 free ... 2. *Psectrodes*,
  d^1. Palpi extending about four times length of head.   [p. 640.
    a^2. Palpi with a slight upward curve, a rounded
      brush on 3rd joint; costa of fore wing
      excised beyond middle .... 1. *Acrodegmia*,
    b^2. Palpi straight; fore wing with the costa   [p. 640.
      slightly excised beyond middle, the apex
      produced upwards.............................. 5. *Saccopleura*,
    c^2. Palpi straight, the 3rd joint fringed with hair   [p. 644.
      below; fore wing with the costa not excised,
      the outer margin excised below apex and
      angled at middle .... 4. *Macna* (♀),
                                       [p. 642.

D. Palpi downcurved.
  a. Palpi extending about three times length of head.
    a^1. Palpi with the 2nd joint oblique, the 3rd long,
      naked, and downcurved; hind wing with
      vein 8 free .... 63. *Curicta* (♀),
    b^1. Palpi rostriform and evenly curved.   [p. 682.
      a^2. Hind wing with vein 3 present.
        a^3. Fore wing with vein 6 stalked with 7, 8, 9. 62. *Anemosa*, p. 682.
        b^3. Fore wing with vein 6 from the cell.
          a^4. Fore wing with vein 7 given off from
            8 before 9.
            a^5. Fore wing with the outer margin
              evenly curved......... .... 64. *Murgisca*, p. 683.
            b^5. Fore wing with the outer margin
              angled at middle .... 67. *Ocresia*, p. 684.
          b^4. Fore wing with vein 7 given off from 8
            after 9.
            a^5. Fore wing with vein 10 from cell ... .. 61. *Bouchis*, p. 681.
            b^5. Fore wing with vein 10 stalked with
              7, 8, 9.................................... 60. *Pelasgis*, p. 681.
      b^2. Hind wing with vein 3 absent .... 52. *Alpheias*, p. 676.

b. Palpi extending once to twice the length of head.
  *a*¹. Hind wing with veins 4, 5 stalked.
    *a*². Hind wing with vein 2 absent.
      *a*³. Fore wing with vein 2 absent .... 53. *Uliosoma*, p. 676.
      *b*³. Fore wing with vein 2 present.
        *a*¹. Fore wing with vein 7 given off from 8
          after 9 ; 10 absent........................... 54. *Acutia*, p. 677.
        *b*¹. Fore wing with vein 7 given off from 8
          before 9 ; 10 present .... 56. *Caphys*, p. 678.
    *b*². Hind wing with vein 2 present.
      *a*³. Fore wing with vein 3 given off from the
        cell or from close to the cell.
        *a*¹. Palpi thickly scaled.
          *a*⁵. Fore wing with vein 7 given off from   [p. 680.
            8 before 9 ; 10 present .... 58. *Cyclopalpia*,
          *b*⁵. Fore wing with vein 7 given off from   [p. 679.
            8 after 9 ; 10 absent..................... 57. *Tetraschistis*,
          *c*⁵. Fore wing with veins 9, 10, 11 absent. 51. *Blepharocerus*,
        *b*¹. Palpi smoothly scaled ; fore wing with   [p. 675.
          veins 8, 9 absent in male................. 70. *Nachaba*, p. 686.
      *b*³. Fore wing with veins 2, 3 stalked .... 49. *Holoperas*, p. 674.
      *c*³. Fore wing with veins 3, 4 on a long stalk. . 48. *Condylolomia*,
                                        [p. 673.

  *b*¹. Hind wing with veins 4, 5 from cell or 4
    absent.
    *a*². Fore wing with veins 2, 3 stalked............... 50. *Galasa*, p. 674.
    *b*². Fore wing with vein 3 from the cell.
      *a*³. Fore wing with veins 4, 5 stalked.
        *a*⁴. Palpi extending about twice the length
          of head ; fore wing with veins 7 and 10
          from cell, 8 and 11 absent ; hind wing
          with vein 2 present .... 46. *Adenopteryx*,
        *b*¹. Palpi extending about the length of   [p. 672.
          head ; fore wing with veins 7, 8, 9
          stalked, 10 absent ; hind wing with vein
          2 absent ....: 55. *Acallis*, p. 677.
      *b*³. Fore wing with veins 4, 5 from the cell.
        *a*⁴. Fore wing with the outer margin evenly
          curved.
          *a*⁵. Palpi slender, the 2nd joint fringed
            with long hair above; antennæ of
            male with a tuft of hair towards
            extremity .... 71. *Semnia*, p. 687.
          *b*⁵. Palpi slender, the 2nd joint not fringed
           above.
            *a*⁶. Fore wing with vein 3 from angle
              of cell ; antennæ of male ciliated... 73. *Arouva*, p. 689.
            *b*⁶. Fore wing with vein 3 from before
              angle of cell ; antennæ of male
              bipectinate .... 72. *Eurypta*, p. 688.
          *c*⁵. Palpi with a tuft of spatulate scales
            at extremity ; the costa of fore wing
            excised .... 59. *Œctoperodes*,
        *d*⁵. Palpi thickly and roughly scaled.   [p. 680.
          *a*⁶. Hind wing with vein 4 present.
            *a*⁷. Palpi extending about twice the
              length of head.
              *a*⁹. Both wings with veins 4, 5   [p. 689.
                widely separated at origin.. ... 74. *Penthesilea*,
              *b*⁹. Both wings with veins 4, 5 from
                angle of cell .... 66. *Chalinitis*, p. 684.

42*

b^7. Palpi extending about the length
   of head.
   a^8. Fore wing with veins 6, 7 stalked.   43. *Monoloxis*,
   b^8. Fore wing with veins 7, 8 stalked.   [p. 670.
      a^9. Fore wing with the costa
         excised; male with tufts of
         hair on median nervure and
         costa ....   44. *Diloxis*, p. 670.
      b^9. Fore wing with the costa
         straight; male with fovea
         in cell ....   45. *Area*, p. 671.
   b^6. Hind wing with vein 4 absent ......   65. *Streptopalpia*,
b^1. Fore wing with the outer margin angled   [p. 683.
   at middle.   [p. 686.
   a^5. Fore wing with vein 10 present ......   69. *Epitamyra*,
   b^5. Fore wing with vein 10 absent ....   68. *Pachypalpia*,
      [p. 685.

Genus ACRODEGMIA.

Acrodegmia, Rag. Ann. Soc. Ent. Fr. 1890, p. 472.

Palpi porrect, extending about four times length of head with a
slight upward curve, thickly scaled, angled in front of head, the
3rd joint with a rounded brush of hair; frons with a tuft of hair;
antennæ of male almost simple; mid and hind tibiæ and 1st
tarsal joints fringed with long hair. Fore wing of male with a
glandular swelling at base of costa below fringed with long hair,
met by a fringe from median nervure; the costa highly arched at
middle, then excised; the apex acute; the outer margin excised
from apex to vein 4, where it is angled; vein 3 from near angle of
cell; 4, 5 from angle; 6, 7 stalked; 8 absent; 9, 10, 11 free.
Hind wing with the apex produced and acute; veins 3, 4, 5 from
angle of cell; 6, 7 stalked, 7 anastomosing with 8.

Fig. 1.

Acrodegmia pselaphialis, ♂. ½.

Type. ACRODEGMIA PSELAPHIALIS, Rag. Ann. Soc. Ent. Fr. 1890, p. 473,
pl. 7. f. 2.   Surinam; Demerara.

Genus PSECTRODES.

Psectrodes, Rag. Ann. Soc. Ent. Fr. 1890, p. 488.

Palpi porrect, straight, and extending about three times length
of head, fringed with hair above and below, the 3rd joint with a
rounded brush of hair; frons with a large tuft; mid and hind
tibiæ and the 1st tarsal joints slightly fringed with hair. Fore

wing with the costa nearly straight; the apex produced and acute;
the outer margin excurved at middle; male with a fringe of hair
from basal part of costa below met by a fringe of hair from
median nervure; the retinaculum hairy; vein 3 from before angle
of cell; 4, 5 from angle; 6 from upper angle; 7, 8 stalked;
9 absent; 10, 11 free.   Hind wing with veins 3, 4, 5 from angle
of cell; 6, 7 from upper angle; 8 free.

Fig. 2.

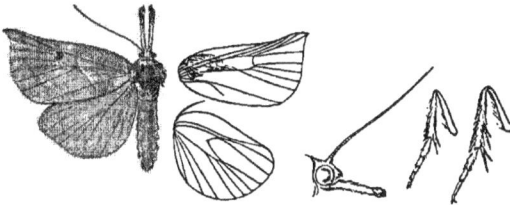

Psectrodes abrasalis, ♂. ¼.

Type. (1)†PSECTRODES ABRASALIS, Wlk. xvi. 39.   Mexico; Brazil.
  Tamyra splendens, Feld. Reis. Nov. pl. 137. f. 15.
  Psectrodes herminialis, Rag. Ann. Soc. Ent. Fr. 1890, p. 488.

(2)†PSECTRODES ILLAPSALIS, Wlk. xvi. 50.   Brazil.

Genus ABÆRA.

Abæra, Wlk. xvi. 76 (1858).

Palpi porrect, straight, and extending about three times the
length of head, fringed with long hair above and below; frons
with a sharp tuft; antennæ of male minutely ciliated; tibiæ
smooth.   Fore wing with the costa usually arched at middle; the
apex rectangular; the outer margin excurved at middle; male
with a small tuft of hair at middle of costa above; a glandular
swelling in cell below covered by fringes of hair from subcostal
and median nervures; vein 3 from near angle of cell; 4, 5 from
angle; 6 from upper angle; 7 absent; 8, 9, 10 stalked; 11 free.
Hind wing with the outer margin slightly angled at vein 2; 3 from
near angle of cell; 6, 7 from upper angle, 7 anastomosing with 8.

Fig. 3.

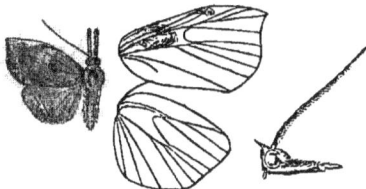

Abæra mactalis, ♂. ¼.

SECT. I. Hind wing with veins 4, 5 well separated at origin.

(1)†ABÆRA MACTALIS, Wlk. xvi. 76.   Brazil.

Sect. II. Hind wing with veins 4, 5 from a point.

(2)†Abæra metallica, n. sp.

♂. Dark brown; palpi ochreous on inner side; hind legs whitish. Fore wing with antemedial whitish line slightly angled below costa and with silvery purple on its inner edge; a postmedial whitish line with silvery purple beyond it, broadest at middle, very much excurved from costa to vein 4, then incurved; an ochreous fascia on apical part of costa; a marginal series of white striæ. Hind wing with submarginal whitish striga, with a small patch of silvery purple on its outer edge above anal angle; a marginal white line.

Hab. Ega, Brazil (*Bates*). *Exp.* 28 mm.

(3)†Abæra rubiginea, n. sp.

♀. Dark red-brown; abdomen fuscous. Fore wing with indistinct antemedial line bent inwards to costa; an indistinct medial line whitish at costa, oblique to vein 6, where it is angled, then sinuous; a prominent postmedial white spot on the costa; traces of a sinuous submarginal series of pale specks; cilia pink, ochreous at tips; costa straight, the outer margin strongly excurved at middle. Hind wing fuscous; the cilia pink.

Hab. Dominica (*W. H. Elliot*). *Exp.* 22 mm.

(4)†Abæra chalcea, n. sp.

♀. Brassy yellow. Fore wing with rufous antemedial line angled below costa; the medial area suffused with pale violet; a postmedial rufous line very obliquely curved from costa to vein 2, then bent outwards to outer angle; outer area suffused with rufous. Hind wing pale fuscous; cilia of both wings pale violet.

Hab. Sta. Martha, Brazil (*Bouchard*). *Exp.* 22 mm.

Genus Macna.

Macna, Wlk. xvi. 78 (1855).
Rhabana, Wlk. xxxiv. 1517 (1865).
Goossensia, Rag. Ann. Soc. Ent. Fr. 1891, p. 97.

Palpi of male upturned to above vertex of head and angled with very long hair in front, of female porrect, straight, extending two to four times length of head, the 2nd joint fringed with hair above and below, the 3rd joint fringed below; frons with a tuft of hair; antennæ of male ciliated; tibiæ and tarsi fringed with long hair. Fore wing with the costa arched; the apex produced to a point; the outer margin excised from apex to vein 4, where it is angled; vein 3 from before angle of cell; 4, 5 from angle; 6 from upper angle; 7, 8, 9 stalked; 11 free. Hind wing with the outer margin very slightly angled at vein 2; the anal angle truncate; vein 3 from close to angle of cell; 4, 5 shortly stalked; 6, 7 stalked, 7 anastomosing slightly with 8. Male with a glandular swelling at base of costa of fore wing below, with a thick oblique tuft of hair from it; the basal half of costa fringed with hair.

Fig. 4.

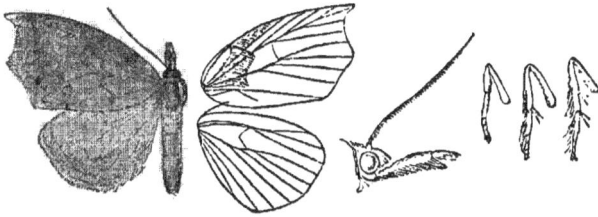

Macna pomalis, ♀. ¼.   Fore wing of ♂.   (From Moths Ind. vol. iv.)

SECT. I. Fore wing with vein 10 stalked with 7, 8, 9 ; palpi of
♀ three to four times the length of head.

Type. (1) MACNA POMALIS, Wlk. xvi. 78.   N.E. India ; Malacca ;
Goossensia prasinalis, Rag. Ann. Soc.   Singapore ; Salanga.
Ent. Fr. 1891, p. 98, & Mon. Phyc. & Gall. pl. 46. f. 4.

(2)†MACNA PLATYCHLORALIS, Wlk. xxxiv. 1517. Andamans ; Java.

SECT. II. Fore wing with vein 10 from the cell ; palpi of
♀ about twice the length of head.

(3)*MACNA ATRIRUFALIS, n. sp.

♂. Head and thorax purplish red-brown ; abdomen fuscous
black.   Fore wing purplish red-brown suffused with fuscous ;
traces of a dark sinuous antemedial line ; two indistinct dark
sinuous postmedial diffused lines excurved at middle, with a pale
speck between them on costa.   Hind wing fuscous brown ; both
wings with white line at base of cilia.   Underside of fore wing
with short oblique white postmedial line from costa ; hind wing
with two dark curved postmedial lines.

♀ with the lines more distinct, the antemedial with a large
fuscous patch inside it on inner area ; the postmedial with fuscous
patch beyond it on costa, the white mark much more prominent ;
a series of black marginal spots.

Hab. Amboina ; Humboldt Bay, N. Guinea.   *Exp.* ♂ 34,
♀ 58 mm.   Type in Coll. Rothschild.

(4)*MACNA IGNEBASALIS, n. sp.

♀. Head and thorax red-brown ; abdomen grey ; metathorax
and base of abdomen tinged with fuscous.   Fore wing pale
purplish red-brown, with a large subbasal patch of fiery orange
and pale yellow scales between cell and vein 1 and with a few
scattered black scales on and near it ; a black fascia on inner
margin ; antemedial line black, straight, and obsolete on costal
area ; a black discocellular spot ; traces of a red postmedial line
excurved at middle ; a straight black submarginal line.   Hind wing
fuscous brown, with reddish marginal band defined by black lines.

Hab. Humboldt Bay, N. Guinea.   *Exp.* 44 mm.   Type in Coll.
Rothschild.

Genus SACCOPLEURA.

Saccopleura, Rag. Ann. Soc. Ent. Fr. 1890, p. 502.

Palpi obliquely porrect, straight, and extending about four times length of head, and fringed with long curved scales; frons with a tuft of scales; antennæ of male ciliated; hind tibiæ with a tuft of hair from base. Fore wing with the costa strongly arched at base, excised beyond middle; the apex produced to a sharp point; the outer margin excurved at middle; male with a tympanic vesicle at base of costa; underside with a large costal fold fringed with hair; the frenulum thickened, flattened and contorted, with a short fork from base; the retinaculum annular; vein 3 from near angle of cell; 4, 5 from angle; 6 from upper angle; 7, 8, 9 stalked; 10, 11 free. Hind wing with the outer margin somewhat angled at middle; vein 3 from before angle of cell, which is produced; 4, 5 from angle; 6, 7 shortly stalked; 8 free.

Fig. 5.

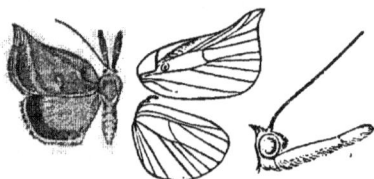

Saccopleura catocalis, ♂. ¼.

Type. *SACCOPLEURA CATOCALIS, Rag. Ann. Soc. Ent. Fr. 1890, p. 503, pl. 7. f. 10.   Chiriqui.

Genus GEPHYRA.

Gephyra, Wlk. xix. 848 (1859).

Palpi porrect, straight, and extending twice to three times the length of head, thickly scaled above and below; frons with a sharp tuft; antennæ of male almost simple; tibiæ nearly smoothly

Fig. 6.

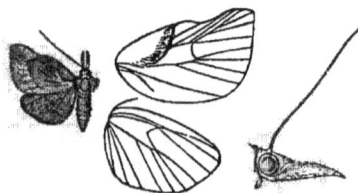

Gephyra getusalis, ♂. ¼.

scaled. Fore wing of male with a large tympanic vesicle at base; the basal half of costa arched, the apical half excised; the costal swelling fringed with hair below; vein 3 from near angle of cell;

4, 5 from angle; 6 from near upper angle; 7, 8, 9 stalked; 10, 11 free. Hind wing with vein 3 from near angle of cell; 4, 5 from angle; 6, 7 from upper angle, 7 anastomosing with 8.

SECT. I. Palpi extending about three times length of head; frenulum thickened.

Type. (1)†GEPHYRA GETUSALIS, Wlk. xix. 849.   Brazil.

(2)*GEPHYRA PUSILLA, Feld. Reis. Nov. pl. 137. f. 11.   Brazil.

SECT. II. Palpi extending about twice the length of head; frenulum normal.

A. Fore wing with a single costal excision.

(3)*GEPHYRA DIFFICILIS, Feld. Reis. Nov. pl. 137. f. 14.   Bogotá.

(4)*GEPHYRA POMPONIUS, Druce, Biol. Centr.-Am., Het. ii. p. 193, pl. 59. f. 24.   Mexico; Guatemala.

B. Fore wing with two costal excisions.

(5)*GEPHYRA CYNISCA, Druce, Biol. Centr.-Am., Het. ii. p. 193, pl. 59. f. 23.   Mexico; Guatemala.

Genus DRYMIARCHA.

Drymiarcha, Meyr. Trans. Ent. Soc. 1885, p. 441.

Palpi porrect, about twice the length of head, thickly scaled and curved towards each other, enclosing the space between; frons with large tuft; antennæ of male with the basal joint dilated and enclosing a hollow, the shaft given off at an angle with a tooth at base and bipectinate with short branches; large paired tufts of hair behind the antennæ; metathorax with paired tufts at origin of hind wing above; tibiæ moderately hairy. Fore wing with the apex somewhat produced and the outer margin excurved; vein 3 from before angle of cell; 4, 5 from angle; 6 from upper angle; 7, 8, 9 stalked; 10 absent; 11 sinuous. Hind wing with vein 3 from near angle of cell; 4, 5 from angle; the discocellulars highly angled; 6, 7 stalked; 8 free.

Fig. 7.

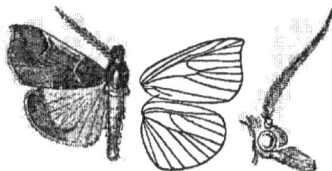

Drymiarcha exanthes, ♂. ½.

Type. *DRYMIARCHA EXANTHES, Meyr. Trans. Ent. Soc. 1885, p. 441.
          Australia.

Genus NEOPHRIDA.

Neophrida, Möschl. Lep. Surinam, p. 26 (1881).

Palpi porrect, straight, extending about twice the length of head and clothed with rough hair ; frons smooth ; antennæ of female nearly simple. Fore wing with the costa highly arched near base, the apex rectangular ; vein 3 from before angle of cell ; 4, 5 from angle, which is much produced ; 6 from upper angle, which is also much produced ; 7, 8, 9 stalked, and 10 approximated to them ; 11 free. Hind wing with vein 3 from close to angle of cell, which is very much produced ; 4, 5 stalked ; 6, 7 stalked ; 8 free.

Fig. 8.

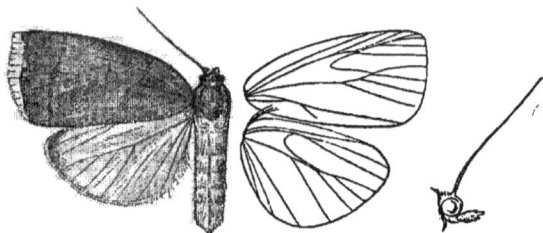

Neophrida aurolimbalis, ♀. ¼.

Type. *NEOPHRIDA AUROLIMBALIS, Möschl. Lep. Surinam, p. 27.
British Guiana ; Surinam.

Genus TAMYRA.

Tamyra, Herr.-Schäff. Samml. aussereur. Schmett. p. 76 (1855).
Lametia, Wlk. xvi. 77 (1858).
Tamyrodes, Rag. Ann. Soc. Ent. Fr. 1890, p. 476.

Palpi upturned, four or five times the length of head and reaching far above it, thickly scaled, the 3rd joint clothed with

Fig. 9.

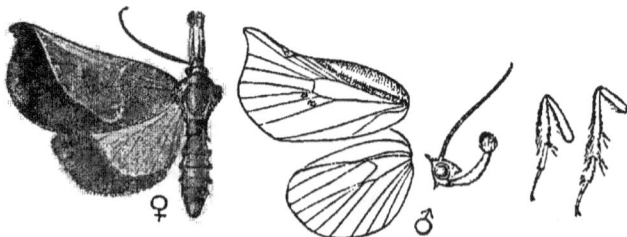

Tamyra ignitalis. ¼.

hair, forming a rounded brush ; frons with a sharp tuft ; antennæ thickened and flattened ; mid and hind tibiæ and 1st joint of hind tarsus above fringed with hair. Fore wing broad, the costa nearly straight in female, excised beyond middle in male and with a small triangular fold; two postmedial tufts of raised scales between veins 2 and 4 and a glandular swelling at base of costa below; the apex slightly produced and acute; the outer margin very much excurved at middle; vein 3 from near angle of cell; 4, 5 from angle; 6 from upper angle; 7, 8, 9 stalked in female, in male 7 from angle and 8, 9, 10 free from before the angle; 11 anastomosing with 12, in female free. Hind wing with the cell short; veins 3, 4, 5 from angle; the discocellulars curved; 6, 7 from upper angle, 7 anastomosing slightly with 8.

Type. (1)*TAMYRA PENICILLANA, Herr.-Schäff. Samml. aussereur. Schmett.
            p. 76, f. 453.   Brazil.

(2)†TAMYRA CUPRINA, n. sp.

♀. Differs from *penicillana* in the fore wing having a diffused red patch below the cell before the antemedial pale line; no discocellular spot: the postmedial oblique line straighter, arising from the costa before the apex and with the whole area beyond it bright chestnut-red. Hind wing with the marginal area reddish.

Hab. Guadaloupe.   *Exp.* 54 mm.

(3)†TAMYRA IGNITALIS, Wlk. xvi. 77.   Brazil.

Auctorum.

Tamyrodes papulalis, Rag. Ann. Soc. Ent. Fr. 1890, p. 475, pl. 7.
            f. 4.   Cayenne.

Genus CASUARIA.

Casuaria, Wlk. xxxv. 1807 (1866).

Palpi upturned, 4 or 5 times the length of head and reaching far above it, the 2nd and 3rd joints roughly scaled and fringed with long hair below; antennæ of male almost simple; tibiæ and 1st joint of hind tarsi fringed with hair above. Fore wing with

Fig. 10.

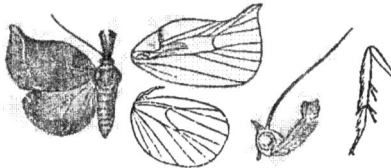

Casuaria armata, ♂ . ¹⁄₁.

tympanic vesicle at base of costa above with ridged membrane across it; the costa highly arched at middle, excised beyond middle; the apex produced upwards and acute; the outer margin rounded; a glandular swelling fringed with hair at base of costa below; the retinaculum annular; the frenulum flattened and twisted near base and with a short lower fork; vein 3 from near angle of cell; 4, 5 from angle; 6 from upper angle; 7, 8, 9 stalked; 10, 11 free. Hind wing with veins 3, 4, 5 from angle of cell; 6, 7 from upper angle, 7 anastomosing slightly with 8.

Type. (1)†CASUARIA ARMATA, Wlk. xxxv. 1807.   New Granada.
Tamyra physophora, Feld. Reis. Nov. pl. 137. f. 10.

(2)*CASUARIA CRUMENA, Feld. Reis. Nov. pl. 137. f. 16.   Bogotá.

Genus RUCUMA.

Rucuma, Wlk. xxviii. 441 (1863).
Erioptycha, Rag. Ann. Soc. Ent. Fr. 1890, p. 496.

Palpi upturned and reaching vertex of head, the 3rd joint minute; frons with a sharp tuft; antennæ of male simple; mid and hind tibiæ and tarsi fringed with long hair above; the tympanic vesicle, retinaculum, and frenulum as in *Casuaria.* Fore wing with the costa very much arched at middle and excised beyond it; the apex produced upwards and falcate; the outer margin much excurved at middle, then excised to anal angle; neuration as in *Casuaria.* Hind wing of male with glandular swelling on basal half of inner margin below, clothed with rough hair.

Fig. 11.

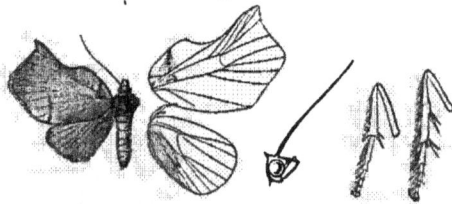

Rucuma recurvana, ♂. ⅓.

Type. †RUCUMA RECURVANA, Wlk. xxviii. 441.   Brazil.

Auctorum.

Erioptycha umbrivittalis, Rag. Ann. Soc. Ent. Fr. 1890, p. 497, pl. 7. f. 9.   Brazil.

Genus ORYCTOPLEURA.

Oryctopleura, Rag. Ann. Soc. Ent. Fr. 1890, p. 495.

Palpi upturned and reaching vertex of head, the 3rd joint minute; frons with a sharp tuft of scales; mid and hind legs thickly clothed with scales. Fore wing with the costa strongly arched at base, then deeply excised; the apex produced to a sharp point; the outer margin excurved at middle; male with a large

tympanic vesicle; a glandular swelling at base of costa below,
fringed with hair; the retinaculum annular; the frenulum
thickened and flattened; veins 2, 3 from well before angle of cell;
4, 5 from angle; 6 from below upper angle; 7, 8, 9, 10 stalked.
Hind wing with the outer margin angled at middle; the costa
arched; the cell short; vein 3 from before angle of cell; 4, 5 from
angle; 6, 7 from upper angle.

Type. *ORYCTOPLEURA ARCUATALIS, Rag. Ann. Soc. Ent. Fr. 1890,
p. 496.   Brazil.

Genus SALOBRENA.

Salobrena, Wlk. xxviii. 446 (1863).
Œctoperia, Zell. Verh. zool.-bot. Ges. Wien, 1875, p. 331.
Clydonopteron, Riley, Ent. Am. iii. p. 287.

Palpi upturned, about twice the length of head and reaching
well above the vertex, fringed with long hair above; frons
rounded; antennæ somewhat annulate; mid legs with large tufts
of scales at middle and end of tibiæ. Fore wing of male with a
tympanic vesicle at base of costa, usually with two excisions
beyond it; the retinaculum annular; the frenulum much
thickened and with a short lower fork; vein 3 from before angle
of cell; 4, 5 from angle; 6 from upper angle; 7, 8, 9 stalked;
10, 11 free. Hind wing with veins 3, 4, 5 from angle of cell;
6, 7 from upper angle, 7 anastomosing with 8.

Fig. 12.

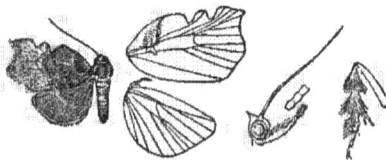

Salobrena excisana, ♂. 3/2.

SECT. I. (*Salobrena*). Mid tarsus of male with a large tuft of scales
on 1st joint; fore wing with two deep excisions in costa.

Type. (1)†SALOBRENA EXCISANA, Wlk. xxviii. 446.   Brazil.
          „   *genualis*, Feld. Reis. Nov. pl. 137. f. 35.

(2)*SALOBRENA CYRISALIS, Druce, Biol. Centr.-Am., Het. ii. p. 192,
pl. 59. f. 20.   Mexico.

(3)*SALOBRENA PROPYLEA, Druce, Biol. Centr.-Am., Het. ii. p. 193,
pl. 59. ff. 21, 22.   Mexico.

(4) SALOBRENA TECOMÆ, Riley, Ent. Am. iii. p. 288, ff. 132, 133.
                  U.S.A.; W. Indies; Brazil; Buenos Ayres.

(5)*SALOBRENA GIBBOSA, Feld. Reis. Nov. pl. 137. f. 36.   Bogotá.

SECT. II. (*Œctoperia*). Mid tarsus of male without the tuft of
scales; fore wing with the excisions in costa slight.

(6) SALOBRENA SINCERA, Zell. Verh. zool.-bot. Ges. Wien, 1873,
    p. 331, pl. x. f. 45.   U.S.A.

SECT. III. Male with no excisions on costa of fore wing; the
glandular swelling on underside very large, with a thick tuft
of hair from its extremity in end of cell.

(7)†SALOBRENA VACUANA, Wlk. xxviii. 441.   W. Indies; Brazil.

Genus IDNEA.

Idnea, Herr.-Schäff. Samml. aussereur. Schmett. p. 75 (1855).
Uzeda, Wlk. xxviii. 442 (1863).
Corybissa, Wlk. xxviii. 445.
Auchoteles, Zell. Hor. Soc. Ent. Ross. 1877, p. 83.

Palpi upturned, reaching vertex of head, the 3rd joint minute;
frons with a tuft of hair; antennæ minutely ciliated; mid and
hind tibiæ and the 1st joint of tarsi fringed with rough hair. Fore
wing with the costa arched at base, excised beyond middle; the
apex produced and falcate; the outer margin very much excurved
at middle; a glandular swelling in male at base of costa below,
with an oblique tuft of hair from it met by a fringe on median
nervure continued for a short way along vein 2; a hyaline patch
beyond the cell; vein 3 from near angle of cell; 4, 5 from angle;
6, 7 from upper angle in male, 8, 9 stalked, 10 free, 11 anastomosing with 12; in female 7, 8, 9 stalked, 10, 11 free. Hind
wing with the outer margin excurved at middle; veins 3, 4, 5
from lower angle of cell (or abnormally 4, 5 on a long stalk); the
discocellulars obliquely curved; veins 6, 7 stalked, 7 anastomosing
with 8.

Fig. 13.

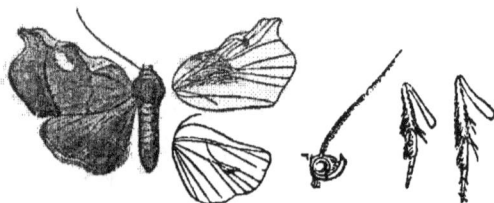

Idnea speculans, ♂. ⅓.

SECT. I. (*Idnea*). Fore wing of male with two excisions in the
costa beyond middle, the lobe between them curled over and
with a tuft of hair; hind wing with short ridges of scales
beyond lower angle of cell on veins 3, 4, 5.

Type. (1) IDNEA SPECULANS, Herr.-Schäff. Samml. aussereur. Schmett.
    ff. 399, 400.   Brazil.
    †*Uzeda olivaceana,* Wlk. xxviii. 443.

SECT. II. Fore wing of male with one excision in the costa beyond middle and no fringed lobe; hind wing without ridges of scales beyond lower angle of cell.

A. (*Corybissa*). Fore wing of male with a rounded lobe at base of costa.

(2)†IDNEA CONCOLORANA, Wlk. xxviii. 439.   Brazil.
  †*Uzeda torquetana*, Wlk. xxviii. 443.
  †*Auchoteles perforatana*, Zell. Hor. Soc. Eut. Ross. xiii. p. 84.
  †   „   *sobriana*, Zell. Hor. Soc. Ent. Ross. xiii. p. 84.

(3)†IDNEA ALTANA, Wlk. xxviii. 438.   Brazil.
  †*Corybissa congruana*, Wlk. xxviii. 416.

B. (*Uzeda*). Fore wing of male with an angled lobe at base of costa; a flap of scales on inner side of the hyaline patch.

a. Fore wing of male with the costal lobe ending before middle, with no tuft of scales on it or ridge on the postmedial line.

(4)†IDNEA PROPRIANA, Wlk. xxviii. 438.   Brazil.
  †*Uzeda vitriferana*, Wlk. xxviii. 442.

b. Fore wing of male with the costal lobe extending to middle and with a tuft of scales on it; a ridge of scales on the postmedial line.

(5)†IDNEA GIBBOSANA, Wlk. xxviii. 444.   Brazil.

Genus SAMCOVA.

Samcova, Wlk. xxviii. 435 (1863).
Epidelia, Rag. Ann. Soc. Ent. Fr. 1891, p. 100.

Palpi obliquely upturned, about twice the length of head and reaching well above it, the 3rd joint long and naked; frons with a tuft of hair; mid tibiæ very thickly fringed with long scales; the 1st tarsal joint with a large tuft; hind tibiæ fringed with long scales at extremity and with a tuft on 1st tarsal joint. Fore wing

Fig. 14.

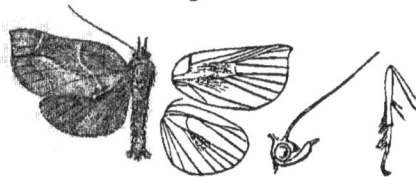

Samcova incensana, ♂. ⅟.

of male with a tympanic vesicle at base of costa; a glandular swelling below, with a fringe of long hair at extremity; a tuft of long hair from median nervure; the costa arched at base; the apex slightly produced and acute; the outer margin excurved at middle; vein 3 from near angle of cell; 4, 5 from angle; 6 from upper angle; 7, 8, 9 stalked; 10, 11 free.   Hind wing with a slight glandular patch of scales beyond cell, which is extremely short; veins 3, 4, 5 from angle; 6, 7 stalked, 7 anastomosing with 8.

Type. (1)†SAMCOVA INCENSANA, Wlk. xxviii. 436.   Brazil.

(2)*SAMCOVA DAMIA, Druce, Biol. Centr.-Am., Het. i. p. 308, pl. 28.
    f. 4 (♀).   Centr. Am.
    Epidelia viridalis, Rag. Ann. Soc. Ent. Fr. 1891, p. 101, pl. 16.
    f. 8.

Genus TOSALE.

Tosale, Wlk. xxviii. 447 (1863).
Fabatana, Wlk. xxxiv. 1265 (1865).
Siparocera, Grote, Ann. N.Y. Lyc. 1876, p. 129.

Palpi upturned, reaching vertex of head, the 3rd joint short; frons with a slight tuft; antennæ of male ciliated; mid tibiæ with large tufts of scales at base and extremity, the first joints of tarsi with large tufts. Fore wing of male with a tympanic vesicle and the costa arched at base; a costal fold below; the retinaculum annulate and fringed with hair; the frenulum thickened and flattened, and with a short lower fork; the apex rounded; the outer angle hooked; vein 3 from near angle of cell; 4, 5 from angle; 6, 7 from upper angle; 8, 9, 10 stalked in male, in female 7, 8, 9, 10 stalked; 11 free. Hind wing with vein 3 from before angle of cell; 4, 5 from angle; the discocellulars highly angled; 6, 7 stalked, 7 anastomosing slightly with 8.

Fig. 15.

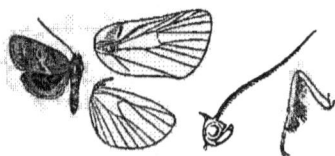

Tosale pyralidoides, ♂ . ⅓.

SECT. I. Fore and hind wings of male without patches of velvety black scales on disk.

(1)†TOSALE AUCTA, n. sp.

♂ . Head, thorax, and abdomen brown with a slight red tinge; palpi and legs deeper red-brown; the extremities of tarsi white. Fore wing red-brown, suffused in parts with grey; a semicircular chocolate band with slightly waved grey outer edge beyond the tympanic vesicle; diffused medial and submarginal olive-brown shades; a grey postmedial line with black specks on its inner edge, excurved from vein 6 to 2, where it is bent inwards; the cilia blackish. Hind wing black-brown, with a pale sinuous line from vein 2 to anal angle. Underside redder, with a dark patch on disk of fore wing ; hind wing with pale curved postmedial line. *Hab.* St. Martha, Brazil. *Exp.* 20 mm.

(2)*TOSALE DECIPIENS, Feld. Reis. Nov. pl. 137. f. 37.   Brazil.

> SECT. II. Male with velvety patches of black scales on disk of fore wing below and on disk of hind wing above.

(3)†TOSALE OVIPLAGALIS, Wlk. xxxiv. 1265.   U.S.A.; Colombia ;
    Siparocera nobilis, Grote, Ann. N.Y. Lyc. 1876, p. 129.   Peru.
    Asopia anthæcioides, Grote, Tr. E. S. Phil. xv. pl. 2. f. 9.

Type. (4)†TOSALE PYRALIDOIDES, Wlk. xxviii. 447.   Brazil.
    Pyralis crassipes, Wlk. xxxiv. 1232.
    Torda metamelana, Wlk. xxxv. 1800.

(5)*TOSALE FLATTALIS, Feld. Reis. Nov. pl. 137. f. 28.   Brazil.

Genus CARCHA.

Carcha, Wlk. xvii. 281 (1859).
Cœloma, Möschl. Lep. Porto Rico, p. 276 (1890).

Palpi upturned and reaching vertex of head, the 3rd joint minute; frons with a tuft of hair; antennæ minutely ciliated ; mid tibiæ with thick tufts of scales at middle and extremity ; hind tibiæ and tarsal joints slightly fringed with hair ; abdomen with medial and paired lateral anal tufts. Fore wing of male with the costa highly arched at base and bearing a tympanic vesicle ; a glandular swelling below ; the retinaculum annulate, the frenulum greatly thickened and flattened, with a strong lower fork ; the disk with a patch of black scales ; the apex rounded ; the outer angle hooked ; veins 2 and 3 from a point before angle of cell ; 4, 5 from angle ; 6 from upper angle ; 7, 8, 9, 10 stalked ; 11 free. Hind wing with vein 3 from before angle of cell ; 4, 5 from angle ; 6, 7 from upper angle, 7 anastomosing with 8.

Fig. 16.

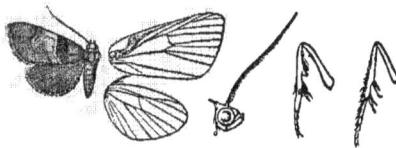

Carcha hersilialis, ♂. ¾.

> SECT. I. Hind wing without ridges of scales on inner area.

Type. (1)†CARCHA HERSILIALIS, Wlk. xvii. 282.   W. Indies; Honduras.
    †*Pyralis dispansalis,* Wlk. xxxiv. 1228.
    †   „   *curtalis,* Wlk. xxxiv. 1230.
    Cœloma tortricalis, Möschl. Lep. Porto Rico, p. 277.
    Tosale moritzi, Rag. Ann. Soc. Ent. Fr. 1890, p. 500.

SECT. II. Hind wing with ridge of large erect scales from below middle of cell to outer margin, thickest towards base.

(2)†CARCHA VIOLALIS, n. sp.

♀. Chocolate-brown with a purple tinge. Fore wing with the costal and apical areas suffused with pink; a marginal series of silvery blue spots. Hind wing with the outer margin and base of cilia silvery blue; a marginal series of black striæ; the scales in the ridge with a metallic tinge. Underside of both wings with indistinct curved postmedial line.

Hab. Espiritu Santo (*Jones*). *Exp.* 24 mm.

Genus SANGUESA.

Sanguesa, Wlk. xxviii. 440 (1863).

Palpi upturned, slender, and reaching just above vertex of head, the 3rd joint minute; frons with a tuft of hair; antennæ of male minutely ciliated; mid and hind tibiæ fringed with long hair, the 1st joint of hind tarsus with a large tuft of hair. Fore wing with a tympanic vesicle at base of costa; the basal half of costa highly arched, then almost straight; the apex rectangular; the outer margin rounded; male with a large tuft of hair from the glandular swelling at base of costa below covering the annular retinaculum; the frenulum very much thickened and flattened, with a short lower fork; a thick tuft of hair below median nervure; vein 3 from before angle of cell; 4, 5 well separated at origin; 6 from upper angle; 7, 8, 9 stalked; 10, 11 free. Hind wing with the costa lobed near base; the outer margin rounded from apex to vein 2, then excised to anal angle; vein 3 from before angle of cell; 4, 5 well separated at origin; the discocellulars highly angled; 6, 7 shortly stalked, 7 anastomosing slightly with 8.

Fig. 17.

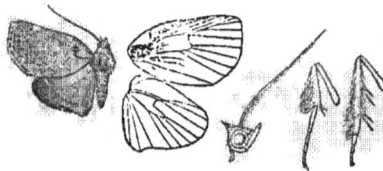

Sanguesa cosmiana, ♂. ⅟.

Type. (1)†SANGUESA COSMIANA, Wlk. xxviii. 440.   Brazil.

(2)†SANGUESA DILATATANA, Wlk. xxviii. 437.   Brazil.

Genus AZAMORA.

Azamora, Wlk. xv. 1757 (1858).
Torda, Wlk. xxviii. 436 (1863).

Amblyura, Led. Wien. ent. Mon. 1863, p. 357.
Thylacophora, Rag. Ann. Soc. Ent. Fr. 1890, p. 490.

Palpi upturned and reaching vertex of head; frons with a sharp tuft; antennæ of male ciliated; hind tibiæ very thickly fringed with long scales, the 1st tarsal joint with a very large tuft; abdomen of male with medial and paired lateral anal tufts. Fore wing of male with a tympanic vesicle at base of costa, which is arched; a glandular swelling below, with a tuft of long hair at extremity met by a fringe from below median nervure; the retinaculum annulate, the frenulum thickened; the apex rectangular or slightly produced; vein 3 from close to angle of cell; 4, 5 from angle; 6 from upper angle; 7, 8, 9 stalked; 10, 11 free. Hind wing with veins 3, 4, 5 from angle of cell; 6, 7 from upper angle; 7 anastomosing slightly with 8.

Fig. 18.

Azamora tortriciformis, ♂. ¼.

SECT. I. Hind tibiæ of male with a tuft of long hair from base; fore wing with a patch of velvety black scales on underside above middle of vein 1, with a tuft of long hair lying over it; hind wing with a patch of velvety black scales in end of cell above.

(1)†AZAMORA MELANOSPILA, Wlk. xxxv. 1799.   Brazil.

SECT. II. Hind tibiæ of male without tuft of hair from base; fore and hind wings without velvety black patches.

A. Fore wing of male with a tuft of pale hair below median nervure on underside.

Type. (2)†AZAMORA TORTRICIFORMIS, Wlk. xv. 1757.   Brazil.
    „   *basiplaga,* Wlk. Trans. Ent. Soc. (3) i. p. 91.
Thylacophora hepaticalis, Rag. Ann. Soc. Ent. Fr. 1890, p. 492.

(3) AZAMORA CORUSCA, Led. Wien. ent. Mon. 1863, p. 357, pl. 6. f. 14.   Brazil.

B. Fore wing of male with a tuft of black hair below median nervure on underside.

(4)†AZAMORA PENICILLANA, Wlk. xxviii. 437.   Brazil.
Thylacophora tortricoidalis, Rag. Ann. Soc. Ent. Fr. 1890, p. 491.

43*

Genus ANISOTHRIX.

Anisothrix, Rag. Ann. Soc. Ent. Fr. 1890, p. 478.

Palpi upturned and reaching vertex of head, the 2nd joint with
a hollow tuft at extremity enclosing the 3rd joint; frons smooth;
antennæ of male with cilia and bristles. Fore wing with the apex
rectangular; vein 3 from near angle of cell; 4, 5 from angle;
6 from upper angle; 7, 8, 9 stalked; 10 free; 11 slightly anastomosing with 12; male with a glandular swelling at base of costa
fringed with hair, which is met by a fringe of hair from median
nervure. Hind wing with veins 3, 4, 5 from angle of cell; the
discocellulars obliquely curved; 6, 7 stalked, 7 anastomosing
with 8.

Fig. 19.

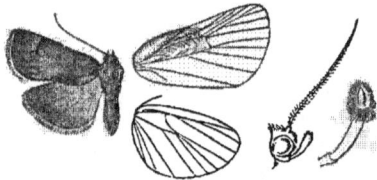

Anisothrix adustalis, ♂ . ¼.

Type. *ANISOTHRIX ADUSTALIS, Rag. Ann. Soc. Ent. Fr. 1890, p. 479,
    pl. 7. f. 5.   Centr. Amer.

Genus HYALOSTICTA, nov.

Palpi upturned, thickly scaled, and reaching vertex of head;
frons smooth; antennæ of male ciliated; mid and hind tibiæ
thickly fringed with long scales, the tarsal joints with large tufts
of scales. Fore wing with the costa slightly arched at base, then
nearly straight; the apex rectangular; male with a slight tuft of
hair at base of costa below; a hyaline fovea in cell; veins 2 and 3
from close to angle of cell; 4, 5 from angle and closely approximated at origin; 6 from near upper angle; 7, 8, 9 stalked; 10, 11
free. Hind wing with veins 3, 4, 5 from angle of cell; 6, 7 from
upper angle, 7 anastomosing with 8.

Fig. 20.

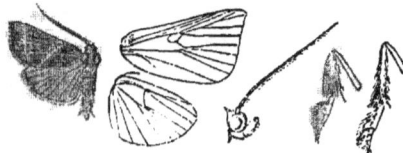

Hyalosticta obliqualis, ♂ . ¼.

Type. †HYALOSTICTA OBLIQUALIS, n. sp.

Head and thorax ochreous brown and purplish fuscous; abdomen
fuscous irrorated with ochreous; legs clothed with reddish,

purplish-black, and ochreous scales. Fore wing purplish fuscous irrorated with greyish ochreous; a very oblique diffused greyish line with ridge on its inner side from costa near base to inner margin beyond middle, the costal area beyond it suffused with grey; an ill-defined submarginal grey line excurved below costa; a reddish patch above outer angle. Hind wing dark fuscous brown. Underside greyer, with dark ante- and postmedial marks on costa of each wing.

Hab. São Paulo (*Jones*).   *Exp.* ♂ 26, ♀ 30 mm.

Genus STHENOBÆA.

Sthenobæa, Rag. Ann. Soc. Ent. Fr. 1890, p. 541.

Palpi upturned, short, smoothly scaled, and not reaching vertex of head; frons smooth; antennæ ciliated; tibiæ smoothly scaled. Fore wing short and broad; the costa slightly arched; the apex rectangular; the outer margin obliquely rounded; vein 2 much curved at origin; lower angle of cell greatly produced by vein 5 running along median nervure; 3, 4, 5 from angle, 4, 5 approximated for some distance; upper angle of cell produced; 6 from well below upper angle; 7 from upper angle; 8, 9 very shortly stalked; 10 from cell; 11 anastomosing with 12. Hind wing with vein 2 from well before angle of cell; 3 and 5 from angle; 4 absent; 6, 7 stalked, 7 anastomosing with 8.

Fig. 21.

Sthenobæa abnormalis, ♂. ⅔.

Type. *STHENOBÆA ABNORMALIS, Rag. Ann. Soc. Ent. Fr. 1890, p. 641.
Ecuador.

Genus HYPOCOSMIA.

Hypocosmia, Rag. Ann. Soc. Ent. Fr. 1890, p. 70.

Palpi upturned, slender, and hardly reaching vertex of head, the 3rd joint minute; frons smooth; antennæ of male thickened; mid tibiæ fringed with long hair, a large tuft on 1st joint of tarsus; hind tibiæ fringed with long hair, the tarsal joints with tufts diminishing distally. Fore wing with the costa nearly straight; the apex rectangular; male with a small tympanic vesicle at base above; the costal thickening slightly fringed with hair below; a fringe of hair from above base of inner margin; the frenulum rather thickened; vein 3 from near angle of cell; 4, 5 from angle; 7, 8, 9, 10 stalked; 11 free. Hind wing with vein 3 from angle of cell; 4, 5 stalked; the discocellulars obliquely curved; 6, 7 from upper angle, 7 anastomosing with 8.

Fig. 22.

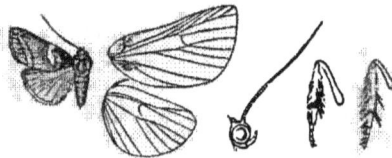

Hypocosmia definitalis, ♂. ¼.

Type. *HYPOCOSMIA DEFINITALIS, Rag. Ann. Soc. Ent. Fr. 1891, p. 505, pl. 7. f. 11.   Venezuela [1].

Genus XANTIPPE.

Xantippe, Rag. Ann. Soc. Ent. Fr. 1890, p. 532.

Palpi upturned, thickly scaled, and hardly reaching vertex of head; frons roughly scaled; antennæ somewhat annulate; mid tibiæ and 1st tarsal joint fringed with hair above; hind tibiæ long, fringed with hair above, the tarsal joints with tufts of scales diminishing distally. Fore wing with the costa straight; the apex rectangular; the outer margin oblique; veins 4, 5 stalked; 6 from upper angle; 7, 8, 9 stalked; 10 from angle; 11 becoming coincident with 12. Hind wing with vein 2 from close to angle of cell; 3, 4, 5 stalked; 6, 7 stalked, 7 anastomosing strongly with 8.

Fig. 23.

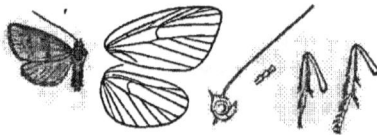

Xantippe auropurpuralis, ♂. ¼.

SECT. I. Fore wing with vein 3 stalked with 4, 5.

Type. (1) XANTIPPE AUROPURPURALIS, Rag. Ann. Soc. Ent. Fr. 1890, p. 533, pl. 5. f. 7.   Brazil.

SECT. II. Fore wing with vein 3 from cell.

(2)†XANTIPPE CHROMALIS, n. sp.

♂. Greenish yellow. Fore wing with the costa fuscous, with pale specks at the origin of the very indistinct pale medial and postmedial lines; a marginal blackish line; the cilia yellowish white, dark at apex. Abdomen and hind wing pale fuscous. Underside fuscous; the costa of fore wing irrorated ochreous and black, of hind wing reddish irrorated with black.

♀ with the hind wing whiter.

Hab. São Paulo (*Jones*). *Exp.* ♂ 18, ♀ 20 mm.

[1] Not Ceylon.

(3) XANTIPPE BICHORDALIS, Rag. Ann. Soc. Ent. Fr. 1890, p. 537.

Brazil.

†*Arta rubricalis*, Warr. A. M. N. H. (6) vii. p. 498.

Genus PARACHMA.

Parachma, Wlk. xxxiv. 1263 (1865).
Zazaca, Wlk. xxxiv. 1269.
Perseis, Rag. Ann. Soc. Ent. Fr. 1890, p. 538.

Palpi upturned, thickly scaled, and hardly reaching vertex of head; frons rounded; antennæ of male ciliated; mid tibiæ with large tufts of scales at middle and extremity and a large tuft on the 1st tarsal joint; hind tibiæ very long in male and roughly scaled, the tarsal joints with tufts of scales diminishing distally; the spurs short. Fore wing with the apex rectangular; veins 4, 5 stalked; 6 from upper angle; 7 absent; 8, 9, 10 stalked; 11 free. Hind wing with vein 2 from near angle of cell; 3 absent; 4, 5 stalked; 6, 7 from upper angle, 7 anastomosing strongly with 8.

Fig. 24.

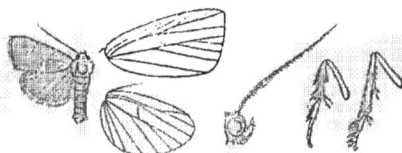

Parachma ochraccalis, ♂. ⅓.

SECT. I. Fore wing with vein 3 stalked with 4, 5.

Type. (1)†PARACHMA OCHRACEALIS, Wlk. xxxiv. 1263.   U.S.A.
  †*Zazaca auratalis*, Wlk. xxxiv. 1269.
  Asopia culiculalis, Hulst, Tr. Am. Ent. Soc. xiii. 167.

(2)†PARACHMA LUTEALIS, n. sp.

Head and thorax yellow marked with fuscous; abdomen yellow suffused with rufous above; the tufts on hind tarsi rufous. Fore wing yellow irrorated with fuscous and red scales; a broad medial purplish-grey band with red edges indented in cell and toothed just below it; the area near outer angle suffused with purplish grey; a marginal red line. Hind wing pale, with some reddish suffusion near lower angle of cell and on outer margin; traces of a postmedial line.

Hab. São Paulo (*Jones*). *Exp.* ♂ 16, ♀ 16–20 mm.

SECT. II. Fore wing with vein 3 absent.

(3)†PARACHMA METERYTHRA, n. sp.

Purplish grey; head and patagia marked with yellowish white; palpi and tarsi ringed with yellowish white. Fore wing with yellowish spot at base of costa and larger triangular medial and postmedial spots, with series of specks arising from them. Hind wing orange-red ; the cilia of both wings pale at tips.

Hab. Espiritu Santo. *Exp.* ♂ 20, ♀ 22 mm.

Genus DASYCNEMIA.

Dasycnemia, Rag. Ann. Soc. Ent. Fr. 1890, p. 489.

Palpi upturned, slender, and not reaching vertex of head, the 3rd joint short ; frons smooth ; antennæ of male finely ciliated ; mid tibiæ with large tufts of scales at middle and extremity, the tarsal joints with tufts of scales diminishing distally ; hind tibiæ long and roughly scaled, the tarsal joints with tufts of scales diminishing distally ; the spurs short. Fore wing with the apex rounded ; vein 3 from angle of cell ; 4, 5 on a long stalk ; 6 from below upper angle ; 7, 8, 9 stalked ; 10, 11 free. Hind wing with vein 3 from angle of cell and closely approximated to 4, 5, which are shortly stalked ; 6, 7 stalked, 7 slightly anastomosing with 8.

Fig. 25.

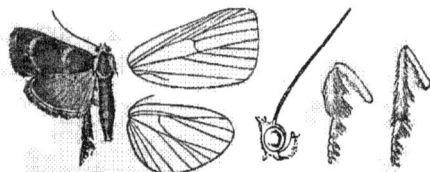

Dasycnemia depressalis, ♂. ¹⁄₁.

Type. *DASYCNEMIA DEPRESSALIS, Rag. Ann. Soc. Ent. Fr. 1890, p. 490, pl. 7. f. 7.   Peru.

Genus CATADUPA.

Catadupa, Wlk. xxviii. 444 (1863).

¹ Palpi with the 2nd joint upturned and reaching vertex of head, the 3rd short and porrect ; antennæ of male ciliated ; hind tibiæ with a fringe of very long scales at extremity ; hind tibiæ long, fringed with rough scales at extremity, the 1st joint of tarsus with a large tuft of scales. Fore wing of male with a large lobe near base of costa, roughly scaled on the costa ; a large ridge of scales on median nervure above, with a hollow between it and the costal lobe ; tufts of hair from base of costa and median nervure below and a large fovea in the cell ; the cell very short ; vein 2 from near angle ; 3, 4, 5 from angle ; 6 from upper angle ; 7, 8, 9 stalked ; 10, 11 free. Hind wing with the cell short ; veins 3, 4, 5 from angle ; 6, 7 from upper angle, 7 anastomosing with 8.

¹ In the unique type the head is wanting, and the characters are taken from Walker's description.

Fig. 26.

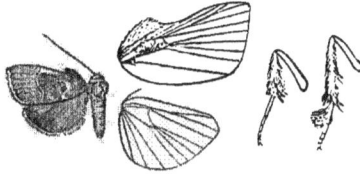

Catadupa integrana, ♂. ⅔.

Type. †CATADUPA INTEGRANA, Wlk. xxviii. 445. Brazil.

Genus DASTIRA.

Dastira, Wlk. xix. 917 (1859).

Palpi upturned, minute and not reaching vertex of head; frons with a sharp tuft; antennæ of male with long curved cilia; mid and hind tibiæ and tarsi nearly smooth. Fore wing with the costa arched at base, the apex rounded; male with a glandular swelling at base of costa below fringed with long hair; vein 3 from near angle of cell; 4, 5 from angle; 6 from upper angle; 7, 8, 9 stalked; 10, 11 free. Hind wing with vein 3 from near angle of cell; 4, 5 from angle; 6, 7 from upper angle, 7 anastomosing with 8.

Fig. 27.

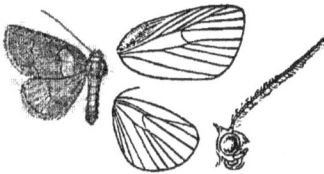

Dastira hippialis, ♂. ⅔.

Type. †DASTIRA HIPPIALIS, Wlk. xix. 917. Brazil.

Genus PROTRICHIA, nov.

Palpi upturned and reaching vertex of head, the 3rd joint minute; frons with a sharp tuft; antennæ of male somewhat thickened; mid and hind tibiæ roughly scaled. Fore wing with the costa slightly excised and with a fringe of hair below it beyond middle; a tuft of hair from base of costa below; veins 3, 4, 5 from close to angle of cell; 6, 7 from upper angle and 8, 9 stalked in male, in female 7, 8, 9 stalked; 10, 11 free. Hind wing with the lower angle of cell produced and the discocellulars obliquely curved; veins 3, 4, 5 from angle; 6, 7 from upper angle, 7 anastomosing with 8.

Fig. 28.

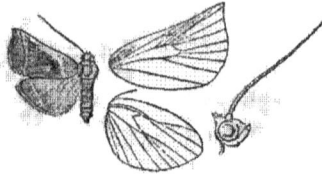

Protrichia vinacea, ♂. ⅟₁.

Type. †PROTRICHIA VINACEA, n. sp.

♂. Head purplish red; thorax and abdomen pale grey-brown. Fore wing with the base of costa purplish red; the basal area grey-brown bounded by an oblique whitish line; the costal area grey-brown; the rest of the wing purplish red, with a slightly sinuous whitish postmedial line. Hind wing pale; the outer area suffused with purplish red; a pale submarginal line angled outwards to the margin at vein 2.

♀ with only a slight vinous tinge on medial area of fore wing; the submarginal line more curved and the area beyond it dark purplish red.

Hab. São Paulo (*Jones*). *Exp.* 24 mm.

Genus SCHISTONEURA.

Schistoneura, Rag. Ann. Soc. Ent. Fr. 1890, p. 527.

Palpi upturned, slender, and reaching vertex of head; frons with a tuft of hair; antennæ of male minutely ciliated; tibiæ thickly scaled. Fore wing of male with the membrane contorted beyond middle below the costa and with tufts of scales on upperside; a large costal fold below, with oblique tuft of hair at its extremity met by a fringe from median nervure; vein 3 from near angle of cell; 4, 5 from angle; 7, 8, 9 stalked in female, 10 free and 11 anastomosing with 12; in male 7 and 8 from upper angle of cell, 9, 10, 11 free and bent up to the costa. Hind wing with the outer margin angled at vein 2; veins 3, 4, 5 from angle of cell; the discocellulars obliquely curved; 6, 7 from upper angle, 7 anastomosing with 8.

Fig. 29.

Schistoneura helicalis, ♂. ⅟₁.

Type. †SCHISTONEURA HELICALIS, Wlk. xviii. 630; Rag. Ann. Soc. Ent. Fr. 1891, pl. 16. f. 2.   Brazil.
    „   *flavitinctalis*, Rag. Ann. Soc. Ent. Fr. 1890, p. 527.

Genus ACROPTERYX.

Acropteryx, Rag. Ann. Soc. Ent. Fr. 1890, p. 469.

Palpi upturned, the second joint reaching vertex of head, the 3rd short in male, long and oblique in female; frons with a tuft of hair; antennæ of male minutely ciliated; tibiæ roughly scaled. Fore wing of male with the costa arched near base, excised beyond middle, the apex slightly produced upwards and falcate; a glandular swelling at base of costa below, with a large oblique tuft of hair at end of it met by a fringe from below median nervure; the costa nearly straight in female; vein 3 from near angle of cell; 4, 5 from angle; 6, 7 from upper angle; 8, 9 absent in male, stalked with 7 in female; 10, 11 free. Hind wing in male with small tufts of scales on veins 5, 6, 7 beyond the cell on underside; vein 3 from near angle of cell; 4, 5 from angle; 6, 7 from upper angle, 7 anastomosing slightly with 8.

Fig. 30.

Acropteryx herbacealis, ♂ . ¼.

SECT. I. Fore wing with the costa excised beyond middle, the apex slightly produced upwards and acute.

(1) ACROPTERYX ARNEA, Cram. Pap. Exot. i. pl. 36, G.

Surinam; Amazons.

Botys linalis, Feld. Reis. Nov. pl. 137. f. 9.

Type. (2) ACROPTERYX HERBACEALIS, Rag. Ann. Soc. Ent. Fr. 1890, p. 470, pl. 7. f. 1.   Chiriqui; St. Martha.

SECT. II. Fore wing with the costa evenly arched; the apex rounded.

(3)*ACROPTERYX NATTERERI, Feld. Reis. Nov. pl. 136. f. 29.

Brazil.

Genus TERATOMORPHA.

Teratomorpha, De Nicéville, Journ. Bom. Nat. Hist. Soc. x. p. 192 (1896).

Palpi upturned, the 2nd joint reaching vertex of head and moderately scaled in front, the 3rd short and naked; frons rounded; antennæ almost simple; tibiæ with the outer spurs about two-thirds length of inner. Fore wing with the costa arched at base and excised beyond middle, the apex bent upwards, arched, and falcate; the outer margin excised below apex, produced and hooked at middle, then excised to outer angle; the inner margin lobed; male with a glandular swelling at base of costa below fringed with tufts of long hair; vein 3 from before angle of cell; 4, 5 from angle : 6 from upper angle; 7, 8, 9 curved and stalked; 10, 11 free. Hind wing with the outer margin somewhat excised below apex and angled at vein 2; vein 3 from near angle of cell; 4, 5 from angle; 6, 7 from upper angle, 7 anastomosing slightly with 8.

Type. †TERATOMORPHA HAMPSONI, De Nicéville, Journ. Bom. Nat. Hist. Soc. x. p. 192, pl. 1. f. 52.   Tenasserim.

Fig. 31.

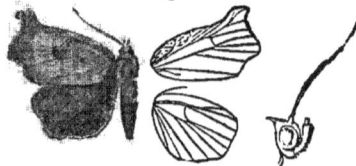

Teratomorpha hampsoni, ♂. ⅓. (From Moths Ind. vol. iv.)

Auctorum.

Goossensia darabitalis, Snell. Tijd. v. Ent. xxxviii. p. 107, pl. 5. f. 1.   Java.

'Genus PIONIDIA, nov.

Palpi upturned, slender, reaching vertex of head and held well in front of frons, which is smooth; antennæ of male almost simple; mid tibiæ thickly fringed with scales; hind tibiæ with a slight tuft of hair from base. Fore wing with the costa arched at base, then straight, the apex rectangular; male with some rough scales in cell below; the cell short and narrow; veins 2, 3 from near angle; 5 from above angle; 6 from below upper angle; 7, 8, 9 stalked; 10, 11 absent. Hind wing with the cell short; vein 2 absent; 3, 4, 5 from angle; the discocellulars highly angled; 6, 7 from upper angle, 7 anastomosing with 8.

Fig. 32.

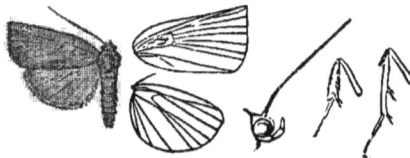

Pionidia albicilia, ♂. ⅓.

Type. †PIONIDIA ALBICILIA, n. sp.

♂. Yellowish brown. Fore wing slightly suffused with pink; a

small tuft of scales on discocellulars; a dark marginal line; the
cilia pure white. Hind wing pale ochreous; the apical area
suffused with pink. Underside of fore wing with pink suffusion
in and beyond cell to outer margin.

Hab. Rio Janeiro. *Exp.* 28 mm.

Genus PARIDNEA.

Paridnea, Rag. Ann. Soc. Ent. Fr. 1891, p. 602.

Palpi upturned, extending far in front of frons and reaching
vertex of head, roughly scaled on outer side, 3rd joint with the
scales directed downwards ; frons nearly smooth ; antennæ strongly
ciliated ; tibiæ nearly smooth. Fore wing with the costa nearly
straight, the apex rectangular ; male with a large tuft of scales on
costa beyond middle ; a glandular swelling at base of costa below
fringed with long hair, met by a fringe from median nervure;
veins 3, 4, 5 from angle of cell : male with veins 6, 7, 8 free from
close to upper angle ; 9, 10 stalked ; 11 free. Hind wing with
veins 3, 4, 5 from angle of cell ; the discocellulars highly angled ;
6, 7 shortly stalked, 7 anastomosing with 8.

Fig. 33.

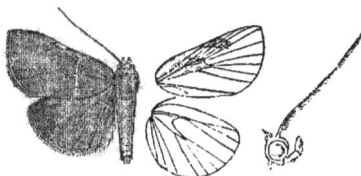

Paridnea holophœalis, ♂. ½.

Type. PARIDNEA HOLOPHÆALIS, Rag. Ann. Soc. Ent. Fr. 1891, p. 603.
Centr. & S. Am.
Stemmatophora demonica, Druce, Biol. Centr.-Am., Het. ii. p. 200,
pl. 60. f. 9.

Genus DERBETA.

Derbeta, Wlk. xxxiv. 1147 (1865).

Palpi upturned and reaching halfway to vertex of head, thickly
scaled, and the 3rd joint minute ; frons with a tuft of hair;
antennæ of male bipectinate (mid and hind legs wanting). Fore
wing long and narrow ; the costa evenly arched; the apex rectangular ; the outer margin obliquely curved ; the inner margin evenly
arched ; a small glandular swelling at base of costa below; the
retinaculum tufted with hair ; veins 3, 4, 5 from angle of cell ; 6
from upper angle ; 7, 8 stalked ; 9, 10 absent ; 11 free. Hind
wing with vein 3 from near angle of cell ; 4, 5 approximated for
a short distance ; the discocellulars highly angled ; 6, 7 from
upper angle, 7 anastomosing slightly with 8.

Fig. 34.

Derbeta nigrifimbria, ♂ . ⅓.

Type. †DERBETA NIGRIFIMBRIA, Wlk. xxxiv. 1148.   Brazil.

Genus PYRAUSTODES.

Pyraustodes, Rag. Ann. Soc. Ent. Fr. 1890, p. 484.

Palpi upturned, slender, very short and hardly reaching half-way to vertex of head; frons rounded ; antennæ almost simple ; mid and hind tibiæ slightly fringed with hair.   Fore wing of male with a large flap of scales in the cell ; the basal half of costal arched ; the apex slightly produced ; the outer margin obliquely rounded ; veins 3, 4, 5 from angle of cell ; 6 from below upper angle ; 7 absent ; 8, 9 stalked ; 10 free ; 11 curved.   Hind wing with the discocellulars very highly angled ; the lower angle of cell greatly produced ; veins 3, 4, 5 from angle ; 6, 7 shortly stalked, 7 anastomosing slightly with 8.

Fig. 35.

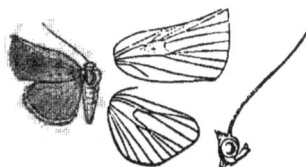

Pyraustodes flavicostalis, ♂ . ⅓.

Type. *PYRAUSTODES FLAVICOSTALIS, Rag. Ann. Soc. Ent. Fr. 1890, p. 485.   Brazil.

Genus ARICA.

Arica, Wlk. xxviii. 439 (1863).

Palpi with the 2nd joint upturned and reaching vertex of head, the 3rd porrect and thickly scaled ; frons with a sharp tuft; antennæ of male nearly simple ; mid tibiæ somewhat thickly clothed with hair ; hind tibiæ long.   Fore wing very broad ; male with tympanic vesicle at base of costa, which is very highly arched, the outer half somewhat excised ; the apex rectangular ; the outer and inner margins forming an almost continuous curve; the retinaculum annulate ; the frenulum thickened and flattened, with a short lower fork ; vein 3 from near angle of cell ; 4, 5 from

angle ; 6 from upper angle ; 7, 8, 9 stalked; 10, 11 free. Hind
wing with veins 3 and 5 well separated from 4; the discocellulars
very highly angled; 6, 7 from upper angle, 7 anastomosing slightly
with 8.

Fig. 36.

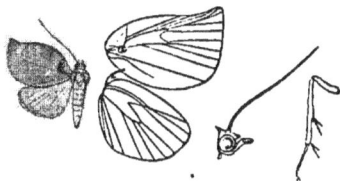

Arica pelopsana, ♂ . }.

Type. (1)†ARICA PELOPSANA, Wlk. xxviii. 439.   Brazil.

(2) ARICA SPLENDENS, Druce, Biol. Centr.-Am., Het. ii. p. 194,
    pl. 59. f. 25.   Panama ; Brazil.

Genus ITAMBE.

Itambe, Rag. Ann. Soc. Ent. Fr. 1891, p. 607.

Palpi minute, porrect, and not reaching nearly to end of frons,
which is rounded and smooth ; antennæ of male minutely ciliated ;
tibiæ nearly smoothly scaled. Fore wing with the basal half of
costa extremely highly arched : the outer half deeply excised ; the
apex produced upwards and falcate ; the outer margin excised
below apex, then excurved ; vein 2 from near angle of cell ; 3, 4, 5
from angle ; 6 from below upper angle ; 7, 8 stalked ; 9 absent ;
10 free ; 11 given off from 12. Hind wing of male with a large
rounded patch of thick scales on upperside occupying the greater
part of inner area ; vein 2 from near angle of cell ; 3 from angle ;
4, 5 stalked ; 6, 7 from upper angle, 7 anastomosing with 8.

Fig. 37.

Itambe fenestalis, ♂ . }.

Type. ITAMBE FENESTALIS, Rag. Ann. Soc. Ent. Fr. 1891, p. 608.   Brazil.

Genus MICROZANCLA, nov.

Differs from *Zanclodes* in the fore wing having the costa slightly
arched at base and much less produced upwards at apex ; male
with a very small glandular swelling at base of costa ; veins 4, 5

stalked; 7, 8, 9 stalked; 11 free. Hind wing with veins 4, 5 stalked.

Fig. 38.

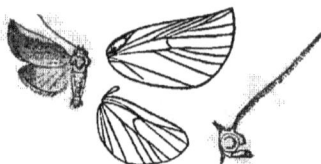

Macrozancla ignitalis, ♂ . ⅓.

Type. †MICROZANCLA IGNITALIS, n. sp.

♂ . Head, thorax, and abdomen brick-red. Fore wing with the basal area brick-red; the outer area dark red-brown with a purplish tinge; the costa from before middle to near apex golden yellow, with a fiery red fascia below it; the inner area broadly fiery red. Hind wing fiery red; the apical area fuscous. Underside of fore wing brown, with the costa deep red.

♀ with no brick-red at base of fore wing, the red-brown and fiery red extending to base.

Hab. São Paulo (*Jones*); Rio Janeiro. *Exp.* ♂ 18, ♀ 20 mm.

Genus ZANCLODES.

Zanclodes, Rag. Ann. Soc. Ent. Fr. 1890, p. 475, pl. 7. f. 3.

Palpi porrect, straight, thickly scaled, and not reaching beyond the large frontal tuft; antennæ of male minutely ciliated; tibiæ roughly scaled on outer side. Fore wing with the basal half of costa highly arched, the apical half excised; the apex very much produced upwards; the outer and inner margins evenly curved; male with a large glandular swelling at base of costa, below fringed with a very thick tuft of hair at extremity; the retinaculum hairy; vein 3 from before angle of cell; 4, 5 from angle; 6, 7 stalked; 8, 9 stalked; 10 free; 11 anastomosing with 12. Hind wing with vein 3 from before angle of cell; 4, 5 from angle; 6, 7 from upper angle, 7 anastomosing with 8.

Fig. 39.

Zanclodes falculalis, ♂ . ⅓.

Type. ZANCLODES FALCULALIS, Rag. Ann. Soc. Ent. Fr. 1890, p. 475,
pl. 7. f. 3.   Brazil.

Genus ARTA.

Arta, Grote, Bull. Buff. Soc. ii. p. 229 (1875).
Heliades, Rag. Ann. Soc. Ent. Fr. 1890, p. 534.

Palpi porrect, straight, and hardly extending beyond the frons, which is rounded; antennæ of male annulate and ciliated; mid and hind tibiæ slightly fringed with hair on outer side, the tarsal joints smooth. Fore wing with the costa nearly straight, the apex rectangular; vein 3 from angle of cell; 4, 5 stalked; 6 from upper angle; 7, 8, 9 stalked; 10 absent; 11 free. Hind wing with vein 2 from angle of cell, which is short; 3, 4, 5 stalked; 6, 7 stalked, 7 anastomosing with 8.

Fig. 40.

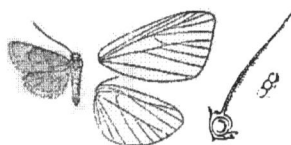

Arta statalis, ♂. ⅓.

(1)†ARTA SERIALIS, n. sp.

Brownish flesh-colour; abdomen yellower, the anal tuft bright ochreous. Fore wing with the inner and medial areas slightly suffused with pink; an oblique medial fuscous line very slightly angled on median nervure; a nearly erect postmedial line; a prominent marginal series of black striæ; the tips of cilia blackish; Hind wing whitish; the apical area tinged with brown; an indistinct postmedial line from costa to vein 5; a series of black marginal striæ. Underside with the costal area of both wings suffused with pink; hind wing with prominent black discocellular spot and postmedial line.

Hab. São Paulo (*Jones*).   *Exp.* 20 mm.

Type. (2)†ARTA STATALIS, Grote, Bull. Buff. Soc. ii. p. 230.   U.S.A.
    „ *epicœnalis*, Rag. Ann. Soc. Ent. Fr. 1890, p. 536.
    „ *mulleolella*, Hulst, Entom. Am. 1887, p. 133.

(3)†ARTA OLIVALIS, Grote, Can. Ent. x. p. 23.   U.S.A.

(4)*ARTA ENCAUSTALIS, Rag. Ann. Soc. Ent. Fr. 1890, p. 537,
    pl. 5. f. 8.   Brazil.

Genus SARCISTIS, nov.

Palpi porrect, thickly scaled, straight and extending rather beyond the frons, which is smoothly scaled; antennæ of female ciliated; tibiæ and tarsi smoothly scaled. Fore wing with the costa almost straight; the apex rounded; vein 2 from near angle of cell; 3, 4, 5 stalked; 6 from near upper angle; 7, 8 stalked; 9 absent: 10 free; 11 becoming coincident with 12. Hind wing with vein 2 from near angle of cell; 3 absent; 4, 5 stalked; 6, 7 from upper angle, 7 anastomosing with 8.

Fig. 41.

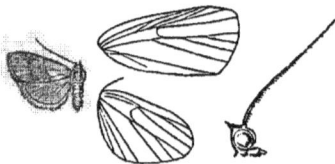

Sarcistis medialis, ♂. ½.

Type. †SARCISTIS MEDIALIS, n. sp.

♀. Flesh-colour. Fore wing thickly irrorated with pink : the costa tinged with yellow; ante- and postmedial very slightly curved fuscous lines, the area between them slightly darker. Hind wing pale with fuscous irroration; a slightly dark marginal line. Underside with pink irroration on fore wing and costal area of hind wing.

Hab. São Paulo (*Jones*). *Exp.* 18 mm.

Genus MONOLOXIS, nov.

Palpi rostriform, downcurved, roughly scaled, and extending about the length of head; frons with a sharp tuft; antennæ of male minutely ciliated; legs smoothly scaled. Fore wing with the costa evenly arched; the apex rectangular; male with a small tuft of hair on costa beyond middle; a glandular swelling at base of costa below fringed with hair, met by a fringe of hair from median nervure; vein 3 from near angle of cell; 4, 5 from angle; 6, 7 stalked; 9 absent; 10, 11 free. Hind wing with vein 3 from near angle of cell; 4, 5 from angle; 6, 7 from upper angle, 7 anastomosing with 8.

Fig. 42.

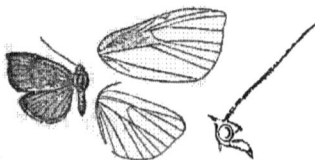

Monoloxis cinerascens, ♂. ½.

Type. †MONOLOXIS CINERASCENS, Warr. A. M. N. H. (6) vii. p. 424.

Brazil.

Genus DILOXIS, nov.

Palpi rostriform, downcurved, thickly scaled, and extending about the length of head; frons smooth; antennæ of male

minutely ciliated ; legs smoothly scaled.   Fore wing with the
basal half of costa arched, the outer half excised ; apex somewhat
produced ; the outer margin obliquely curved ; male with a large
tuft of scales on median nervure above, and a smaller tuft from
costa beyond middle ; veins 3, 4, 5 from angle of cell; 6 from upper
angle ; 7, 8 stalked ; 9 absent ; 10, 11 free.   Hind wing with veins
3, 4, 5 from angle of cell, which is produced ; 6, 7 from upper angle,
7 anastomosing with 8.

Fig. 43.

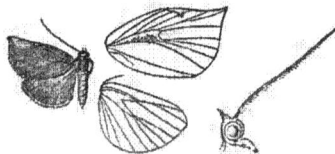

Diloxis ochriplaga, ♂. ½.

Type. †DILOXIS OCHRIPLAGA, n. sp.

♂. Head and collar red-brown ; thorax and abdomen dark
brown.   Fore wing with the costal half red-brown ; the tufts
purplish ; a large ochreous patch on costa before apex ; the inner
half dark brown.   Hind wing dark brown.

Hab. Rio Janeiro.   *Exp.* 20 mm.

Genus AREA.

Area, Rag. Ann. Soc. Ent. Fr. 1890, p. 483.

Palpi rostriform, downcurved, nearly smoothly scaled, and
extending about the length of head ; frons smooth ; antennæ of
male somewhat thickened ; mid and hind tibiæ somewhat hairy.
Fore wing with the costa straight ; the apex rectangular ; male
with a hyaline fovea in end of cell ; vein 3 from near angle of cell ;
4, 5 from angle ; 6 from upper angle ; 7, 8 stalked ; 9 absent ;
10, 11 free in female, 10 absent in male.   Hind with vein 3 from
near angle of cell ; 4, 5 from angle, which is extremely produced ;
6, 7 stalked, 7 anastomosing with 8.

Fig. 44.

Area diaphanalis, ♂. ½.

Type. AREA DIAPHANALIS, Rag. Ann. Soc. Ent. Fr. 1890, p. 484.
Brazil ; Argentina.
44*

Genus ADENOPTERYX.

Adenopteryx, Rag. Ann. Soc. Ent. Fr. 1890, p. 507.

Palpi porrect, downcurved, and extending about twice the length of head; frons with a tuft of scales; antennæ of male ciliated, the basal joint long; hind tibiæ long and slightly hairy. Fore wing with the costa arched; the apex rounded; male with a large glandular swelling at base of costa; vein 3 from angle of cell, 4, 5 shortly stalked; 6 and 7 from below angle of cell; 9, 10 from angle, 10 running almost at right angles to costa; 8 and 11 absent. Hind wing with the cell short, and produced at lower angle; 3 from before angle of cell; 4, 5 from angle; 6, 7 from upper angle, 7 anastomosing with 8.

Fig. 45.

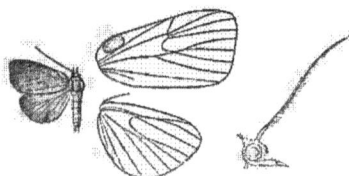

Adenopteryx conchyliatalis, ♂. ⅟₁.

Type. *ADENOPTERYX CONCHYLIATALIS, Rag. Ann. Soc. Ent. Fr. 1890, p. 508, pl. 5. f. 4.   Algeria [1].

Genus HYPERPARACHMA.

Hyperparachma, Warr. A. M. N. H. (6) viii. p. 62 (1891).

Palpi porrect, straight, and hardly extending beyond the frons, the end of 2nd joint thickly scaled; frons roughly scaled; antennæ of male somewhat annulated and ciliated; tibiæ nearly smooth.

Fig. 46.

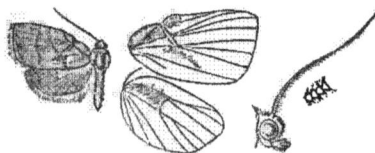

Hyperparachma bursarialis, ♂. ⅔.

Fore wing of male with a glandular swelling at base of costa below fringed with long hair, beyond which on upperside is a deep groove with a flap of scales on its outer edge on costa, and a ridge of

[1] The species is of a typical Neotropical form; the unique type has been kindly lent to me by M. de Joannis, who assures me that it was undoubtedly taken in Philippeville, Algeria. It seems probable that it must in some way have been imported there.

scales below it ; the apex rounded ; the cell very short ; vein 2 absent ; 3, 4, 5 from angle; 6, 7, and 10 from upper angle ; 8, 9, and 11 absent. Hind wing of male with a fringe of long hair in cell below ; vein 2 absent ; 3 from angle ; 4, 5 stalked ; 6, 7 from upper angle, 7 anastomosing with 8.

Type. †HYPERPARACHMA BURSARIALIS, Wlk. xxxiv. 1231.   Honduras ;
                                                         W. Indies ; Brazil.
†   „   *rubrifusca*, Warr. A. M. N. H. (6) viii. p. 62.

Genus CONDYLOLOMIA.

Condylolomia, Grote, Bull. Buff. S. N. S. i. p. 176 (1873).

Palpi rostriform, downcurved, thickly scaled, and extending about the length of head ; frons with a sharp tuft; antennæ of male ciliated ; hind tibæ thickly tufted with scales. Fore wing with the costa evenly curved ; the apex rounded ; male with a tuft of hair on costa before middle ; the cell extremely short ; veins 2, 3, 4 stalked ; 6 from below upper angle ; 9, 10 absent ; 11 free ; female with the cell longer, vein 3 from angle. Hind wing with vein 2 absent ; 3 from angle ; 4, 5 stalked ; 6, 7 from upper angle, 7 anastomosing with 8.

Fig. 47.

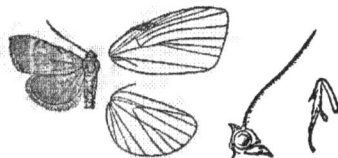

Condylolomia participalis, ♂. ⅔.

SECT. I. Fore wing of male with vein 5 from angle of cell.

Type. (1)†CONDYLOLOMIA PARTICIPALIS, Grote, Bull. Buff. S. N. S. i. p. 177, pl. 5. ff. 4 & 5.   U.S.A.

SECT. II. Fore wing of male with vein 5 stalked with 3, 4 ; a fovea fringed with scales on upperside below the costa before the tuft ; the cell clothed with rough scales ; hind wing with a fringe of very long hair in cell above ; a fringe below the cell and vein 3 ; the inner area thickly clothed with hair, and the anal angle lobed.

(2)†CONDYLOLOMIA METAPACHYS, n. sp.

♂. Head, thorax, and abdomen pale, variegated with red and ochreous. Fore wing pale, suffused with ochreous and red and irrorated with fuscous ; the costal tuft and apex fuscous ; traces of a minutely dentate postmedial white line ; a marginal series of fuscous striæ. Hind wing pale fuscous, the fringes of hair in and below the cell blackish ; the hair on anal lobe rufous.

Hab. São Paulo (*Jones*). *Exp.* 18 mm.

Genus HOLOPERAS.

Holoperas, Warr. A. M. N. H. (6) vii. p. 500 (1891).

Palpi rostriform, downcurved, and extending about the length of
head ; frons smooth ; antennæ of male annulated ; mid femora
and tibiæ with large tufts of scales ; hind legs with tufts of scales
at extremity of tibiæ and on 1st tarsal joint.   Fore wing narrow ;
the costa very highly arched ; the apex rounded ; veins 2, 3 on a
very long stalk from near angle of cell ; 4, 5 shortly stalked ; 6
from close to upper angle ; 7, 8, 9 stalked ; 10, 11 absent ; male
with a costal fold below.   Hind wing with vein 2 from near angle
of cell ; 3 from angle, which is very much produced ; 4, 5 stalked ;
6, 7 from upper angle, 7 anastomosing slightly with 8.

Fig. 48.

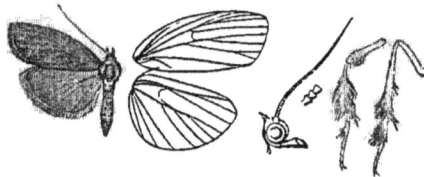

Holoperas innotata, ♂ . ¹⁄₁.

Type. †HOLOPERAS INNOTATA, Warr. A. M. N. H. (6) vii. p. 500.
Colombia.

Auctorum.

Holoperas œnochroalis, Rag. Ann. Soc. Ent. Fr. 1890, p. 509.
Centr. Am.

Genus GALASA.

Galasa, Wlk. xxxv. 1801 (1866).
Cordylopeza, Zell. Verh. z.-b. Ges. Wien, 1873, p. 306.
Palpi rostriform, downcurved, and extending about the length

Fig. 49.

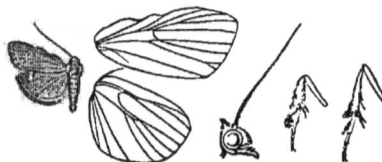

Galasa rubidana, ♂ . ¹⁄₁.

of head ; frons with a sharp tuft ; antennæ of male somewhat
annulate ; mid tibiæ thickly tufted with hair, the 1st tarsal joint

with a large tuft; hind tibiæ tufted with hair at extremity, the
1st tarsal joint with a large tuft. Fore wing with the costa
excised at middle; the apex rounded; veins 2 and 3 on a long
stalk; 4, 5 from angle; 6 from upper angle; 7, 8 stalked; 9 absent;
10, 11 free. Hind wing with veins 3, 4, 5 from angle of cell, which
is very much produced; 6, 7 stalked, 7 anastomosing with 8.

Type. GALASA RUBIDANA, Wlk. xxxv. 1802.   U.S.A.; Jamaica.
Cordylopeza nigrinodis, Zell. Verh. z.-b. Ges. Wien, 1873, p. 306,
  pl. 3. f. 3.
Galasa rubrana, Fitch, Smith List Lep. Bor.-Am. p. 80.
  ,,   *palmipes,* Grote & Rob. Smith List Lep. Bor.-Am. p. 80.

Auctorum.

Galasa daulisalis, Druce, Biol. Centr.-Am., Het. ii. p. 195, pl. 60.
  f. 4.   Panama.

Genus BLEPHAROCERUS.

Blepharocerus, Blanch. Gay's Chili, vii. p. 102 (1852).
Œdematodes, Rag. Ann. Soc. Ent. Fr. 1891, p. 623.

Palpi rostriform, thickly scaled, and extending about the length
of head; frons smooth; antennæ of male annulated and fascicu-
late; mid and hind tibiæ fringed with long hair. Fore wing with
the costa nearly straight, the apex rounded; vein 2 from close to
angle of cell, which is very short; 3, 4 shortly stalked; 5 from
angle; 6 from upper angle; 7, 8 stalked; 9, 10 absent; 11 becoming
coincident with 12, which is very short; female with vein 3 from
near angle of cell; 4, 5 stalked; 7, 8, 9 stalked. Hind wing with
vein 2 from near angle of cell, which is very short, the lower angle
much produced; 3 from angle; 4, 5 stalked; 6, 7 from upper
angle, 7 anastomosing with 8.

Fig. 50.

Blepharocerus chilensis, ♂. ⅓.

SECT. I. (*Blepharocerus*). Fore wing of male with no tuft on costa.

Type. (1) BLEPHAROCERUS ROSELLUS, Blanch. Gay's Chili, vii. p. 102,
  pl. 7. f. 12.   Chili.
    †*Asopia rufulalis,* Led. Wien. ent. Mon. 1863, p. 343, pl. 7.
    f. 3.

SECT. II. (*Œdematodes*). Fore wing of male with a small tuft of
    hair on upperside of costa before middle.

(2)†BLEPHAROCERUS CHILENSIS, Zell. Verh. z.-b. Ges. Wien, xxiv.
    p. 426.   Chili.
    †*Actenia rubescens*, Butl. Tr. Ent. Soc. 1883, p. 51.
    †*Blepharocerus cinerosus*, Warr. A. M. N. II. (6) vii. p. 494.
    †   „   *sabulosus*, Warr. A. M. N. H. (6) vii. p. 495.

Genus ALPHEIAS.

Alpheias, Rag. Ann. Soc. Ent. Fr. 1890, p. 543.

Palpi rostriform, downcurved, slender, and extending about
three times length of head, the 3rd joint minute; frons smooth;
antennæ of female almost simple; tibiæ smoothly scaled.   Fore
wing elongate; the costa arched towards apex, which is rounded;
vein 3 from angle of cell; 4, 5 stalked; 6 from below upper
angle; 7, 8, 9 stalked; 10, 11 free. Hind wing with the cell long;
vein 3 absent; 4, 5 from angle; 6, 7 from upper angle, 7 anastomosing with 8.

Fig. 51.

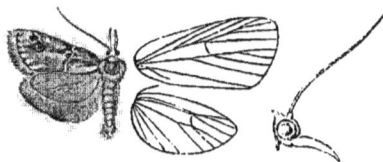

Alpheias baccalis, ♂. ½.

Type. (1)*ALPHEIAS BACCALIS, Rag. Ann. Soc. Ent. Fr. 1890, p. 544,
    pl. 5. f. 11.   Mexico.

(2)*ALPHEIAS GITONALIS, Rag. Ann. Soc. Ent. Fr. 1890, p. 544.
                                                             Mexico.

Genus ULIOSOMA.

Uliosoma, Warr. A. M. N. H. (6) vii. p. 105 (1891).

Palpi rostriform, downcurved, thickly clothed with hair, and

Fig. 52.

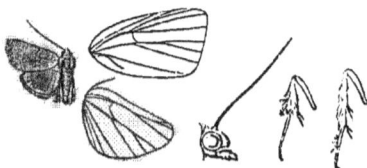

Uliosoma discoloralis, ♂. ½.

reaching just beyond the large frontal tuft; antennæ of male almost simple ; mid tibiæ with a large tuft of scales on outer side; hind tibiæ and 1st tarsal joint fringed with long hair; abdomen with a pair of very large lateral tufts from base.   Fore wing with the costa evenly arched ; the apex rounded ; vein 2 absent ; 3 from near angle of cell ; 4, 5 stalked ; 6, 7 from upper angle; 8, 9 stalked; 10, 11 absent.   Hind wing with vein 2 absent ; 3 from near angle of cell ; 4, 5 stalked : 6 absent ; 7 anastomosing with 8.

Type. †ULIOSOMA DISCOLORALIS, Wlk. xxxiv. 1315.   Brazil.

Genus ACUTIA.

Acutia, Rag. Ann. Soc. Ent. Fr. 1890, p. 539.

Palpi rostriform, downcurved, thickly scaled, and extending about twice the length of head : frons roughly scaled ; antennæ of male strongly ciliated ; hind tibiæ with a slight tuft of hair on outer side from base.   Fore wing narrow ; the apex produced and falcate ; the outer margin very oblique ; vein 2 from close to angle of cell ; 3 from angle; 4, 5 on a long stalk ; 6 from near upper angle ; 7, 8, 9 stalked ; 10 absent ; 11 free.   Hind wing with vein 2 absent : 3 from angle ; 4, 5 on a long stalk ; 6, 7 from upper angle, 7 anastomosing with 8.

Fig. 53.

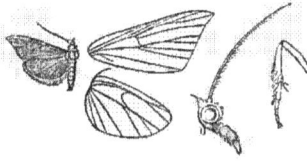

Acutia falciferalis, ♂ .   ⅟.

Type. ACUTIA FALCIFERALIS, Rag. Ann. Soc. Ent. Fr. 1890, p. 540.
                                                                  Brazil.

Genus ACALLIS.

Acallis, Rag. Ann. Soc. Ent. Fr. 1890, p. 540.

Palpi rostriform, downcurved, thickly scaled, and extending about the length of head ; frons roughly scaled ; antennæ of male ciliated ;

Fig. 54.

Acallis fernaldi, ♂ .   ⅟.

tibiæ smoothly scaled.   Fore wing long and narrow; the outer margin oblique ; vein 2 from close to angle of cell ; 3 from angle ;

4, 5 stalked; 6 from upper angle; 7, 8, 9 stalked; 10 absent; 11 free. Hind wing with vein 2 absent; 3 from near angle of cell, which is produced; 4, 5 from angle; 6, 7 stalked, 7 anastomosing with 8.

Type. ACALLIS FERNALDI, Rag. Ann. Soc. Ent. Fr. 1890, p. 540. Colorado.
†*Ugra angustipennis*, Warr. A. M. N. H. (6) vii. p. 494.

Genus CAPHYS.

Caphys, Wlk. xxvii. 13 (1863).
Ugra, Wlk. xxvii. 188.
Euexippe, Rag. Ann. Soc. Ent. Fr. 1890, p. 538.

Palpi rostriform, downcurved, thickly scaled, and extending about the length of head; frons smooth; antennæ of male ciliated; mid tibiæ strongly fringed with scales; hind tibiæ slightly fringed. Fore wing with the costa straight; the apex rectangular; vein 3 from near angle of cell; 4, 5 stalked; 6 from upper angle; 7, 8, 9, 10 stalked; 11 free. Hind wing with vein 2 absent; 3 from angle of cell; 4, 5 stalked; 6, 7 from upper angle, 7 anastomosing with 8.

Fig. 55.

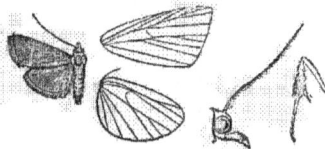

Caphys bilinea, ♂. ¹⁄₁.

SECT. I. Fore wing of male with the costa not indented.

Type. (1)†CAPHYS BILINEA, Wlk. xxvii. 13.   Honduras; W. Indies;
      †*Ugra parallela*, Wlk. xxvii. 188.   Brazil, Venezuela.
      Scopula parallelalis, Wlk. xxxiv. 1462.
      Euexippe bistrialis, Rag. Ann. Soc. Ent. Fr. 1890, p. 539.

(2)†CAPHYS SUBROSEALIS, Wlk. xxxiv. 1462.   Honduras.

(3)†CAPHYS DUBIA, Warr. A. M. N. H. (6) vii. p. 495.
                                                  W. Indies; Brazil.
(4)†CAPHYS PALLIDA, n. sp.

   ♂. Pale greyish ochreous. Fore wing with a slight purplish tinge; the basal half of costa blackish; traces of sinuous ante- and postmedial dark lines; a marginal series of black specks; the cilia fuscous at tips. Hind wing darker and with a reddish tinge; a marginal series of black specks; the cilia fuscous. Underside of fore wing suffused with rufous; the basal half suffused with fuscous; hind wing paler, with rufous and fuscous suffusion on costal and outer areas and with a curved dark postmedial line.
   Hab. São Paulo (*Jones*). *Exp.* 14 mm.

SECT. II. Fore wing of male with a slight indentation and tuft of scales at middle of costa and a fovea in cell below.

(5)†CAPHYS FOVEALIS, n. sp.

♂. Head ochreous white; thorax purplish pink; abdomen pale brown. Fore wing bright purplish pink, with a pale speck in cell and traces of a pale postmedial line excurved below costa; cilia blackish at tips. Hind wing fuscous. Underside with the costal area of each wing purplish irrorated with black.

Hab. São Paulo (*Jones*). *Exp.* 12 mm.

SECT. III. Fore wing of male with two slight indentations in costa.

(6) CAPHYS PALMIPES, Feld. Reis. Nov. pl. 127. f. 23.   Brazil.

Auctorum.

Aglossa gryphalis, Hulst, Tr. Ent. Soc. 1876, p. 146.   U.S.A.

Genus TETRASCHISTIS, nov.

Palpi downcurved, projecting about twice the length of head, and thickly scaled; frons with a sharp tuft; antennæ of female simple; mid tibiæ and tarsi fringed with long hair. Fore wing with the costa nearly straight; the apex rectangular; vein 2 from near angle of cell; 3 from angle; 4, 5 on a long stalk; 6 from upper angle; 7 given off from 8 after 9; 10 absent. Hind wing with vein 2 from near angle of cell, which is produced; 3 from angle; 4, 5 stalked; 6, 7 stalked, 7 anastomosing with 8.

Fig. 56.

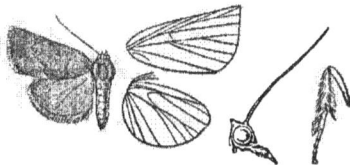

Tetraschistis tinctalis, ♀. ¼.

Type. (1)†TETRASCHISTIS TINCTALIS, n. sp.

♀. Head and thorax rufous; palpi fuscous; abdomen brownish. Fore wing rufous irrorated with fuscous; traces of a sinuous medial dark line arising from a black speck on costa; a discocellular black speck; a postmedial slightly waved fuscous line arising from a black speck on costa and excurved from costa to vein 3; a dark marginal line; cilia fuscous with a pinkish tinge. Hind wing pale, with traces of a curved dark postmedial line more prominent on underside; cilia darkish towards apex.

Hab. São Paulo (*Jones*). *Exp.* 26–30 mm.

(2)†TETRASCHISTIS MAJOR, Warr. A. M. N. H. (6) vii. p. 500.
Colombia.

Genus CYCLOPALPIA, nov.

Palpi thickly scaled, in male curved inwards and downwards and
extending just beyond the large frontal tuft, in female curved downwards and extending about twice the length of head; antennæ of male
serrated ; mid and hind tibiæ moderately fringed with hair. Fore
wing with the costa nearly straight ; the apex acute ; the outer
margin excurved at middle ; vein 3 from near angle of cell ; 4, 5
stalked ; 6 from below upper angle ; 7, 8, 9, 10 stalked ; 11 free.
Hind wing with vein 3 from near angle of cell ; 4, 5 stalked ; 6,
7 shortly stalked, 7 anastomosing with 8.

Fig. 57.

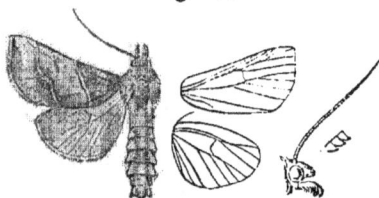

Cyclopalpia violescens, ♂ . ⅓.

Type. †CYCLOPALPIA VIOLESCENS, n. sp.

♂ . Head and, thorax purple suffused with grey ; abdomen
purplish. Fore wing purple thickly suffused with grey, leaving the
basal area and the area on each side of lower part of postmedial
line most purple ; an oblique dark-edged pale antemedial line ; a
dark discocellular spot ; a dark-edged pale postmedial line slightly
bent inwards at vein 4. Hind wing pale suffused with red, especi-
ally on apical area. Underside purplish, the inner half of hind
wing white ; both wings with discocellular lunule ; the apical area
bright chestnut-red, with a white line on its inner edge.
Hab. São Paulo (*Jones*). *Exp.* 40 mm.

Genus ŒCTOPERODES.

Œctoperodes, Rag. Ann. Soc. Ent. Fr. 1891, p. 612.

Palpi rostriform, downcurved, extending about the length of
head, the extremity with a tuft of long scales dilated at tips ; frons
smooth ; antennæ of male minutely ciliated ; mid and hind tibiæ
and first tarsal joints with tufts of scales. Fore wing with the
costa strongly arched at base, then excised ; the apex acute ; the
outer margin oblique towards outer angle ; male with a tympanic
vesicle at base of costa ; a glandular swelling below fringed with
long hair, met by a fringe of hair from median nervure ; vein 3
from near angle of cell ; 4, 5 from angle ; 6, 7 shortly stalked ; 8,
9 stalked ; 10, 11 free. Hind wing with the outer margin angled
at middle ; the lower angle of cell produced ; veins 3 and 5 well
separated from 4 ; 6, 7 shortly stalked.

Type. *ŒCTOPERODES RUFITINCTALIS, Rag. Ann. Soc. Ent. Fr. 1891, p. 613.
Brazil.

Genus PELASGIS.

Pelasgis, Rag. Ann. Soc. Ent. Fr. 1890, p. 487.

Palpi rostriform, downcurved, roughly scaled, and extending three or four times length of head; frons smooth; antennæ of female slightly ciliated; legs smoothly scaled. Fore wing with the costa slightly arched; the apex retangular; vein 3 from near angle of cell; 4, 5 from angle; 6 from below upper angle; 7, 8, 9, 10 stalked, 7 being given off after 9; 11 free. Hind wing with vein 3 from near angle of cell; 4, 5 very shortly stalked; 6, 7 shortly stalked, 7 anastomosing with 8.

Fig. 58.

Pelasgis hypogryphalis, ♀ . ⅟.

Type. *PELASGIS HYPOGRYPHALIS, Rag. Ann. Soc. Ent. Fr. 1890, p. 487.

Brazil.

Genus BONCHIS.

Bonchis, Wlk. Tr. Ent. Soc. (3) i. p. 128 (1862).
Ethnistis, Led. Wien. ent. Mon. 1863, p. 345.
Vurna, Wlk. xxxiv. 1189 (1865).
Zarania, Wlk. xxxiv. 1262.
Gazaca, Wlk. xxxiv. 1273.

Palpi rostriform, strongly downcurved, thickly scaled, and extending about three times length of head; frons smooth; antennæ of male with fascicles of cilia; mid and hind tibiæ with strong tufts of scales. Fore wing of male with the costa produced to a

Fig. 59.

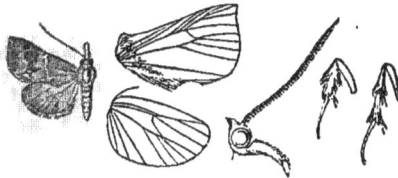

Bonchis scoparioides, ♂ . ⅟.

lobe at base; a thick ridge of scales on basal area and medial and postmedial tufts on inner margin; the retinaculum formed by a tuft of hair from a costal fold; the costa nearly straight; the apex rectangular; vein 3 from near angle of cell; 4, 5 from angle; 6 from near upper angle; 7, 8, 9 stalked: 10, 11 free.   Hind wing

with the median nervure pectinated above; the cell long: vein 3
from near angle : 4, 5 from angle ; 6, 7 from upper angle, 7 anastomosing with 8.

(1) Bonchis munitalis, Led. Wien. ent. Mon. 1863, p. 55, pl. 6,
f. 13.   W. Indies; Honduras.
    †*Vurna instructalis*, Wlk. xxxiv. 1189.
    Zarania cossalis, Wlk. xxxiv. 1262.
    †*Gazaca dirutalis*, Wlk. xxxiv. 1274.

Type. (2) Bonchis scoparioides, Wlk. Trans. Ent. Soc. (3) i. p. 128.
                                          Trinidad; Brazil.

Genus ANEMOSA.

Anemosa, Wlk. xix. 849 (1859).

Palpi rostriform, downcurved at extremity, thickly scaled, and
extending about three times length of head, frons with a sharp
tuft of hair.   Antennæ of male bipectinate, with short branches;
tibiæ thickly scaled.   Fore wing with the apex rounded; vein 3
from near angle of cell ; 4, 5 from angle ; the discocellulars highly
angled; 6, 7, 8, 9 stalked and curved; 10, 11 free.   Hind wing
with vein 3 from before angle of cell ; 4, 5 from angle ; the discocellulars highly angled; 6, 7 stalked, 7 anastomosing with 8.

Fig. 60.

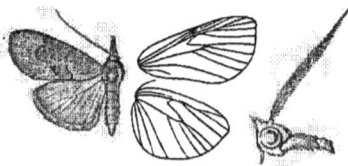

Anemosa isadasalis, ♂.   ¼.

Type. †Anemosa isadasalis, Wlk. xix. 849.   Australia.

Genus CURICTA.

Curicta, Wlk. xxxiv. 1129 (1865).

Palpi of male upturned to vertex of head, the 3rd joint fringed
with long downcurved hair in front, of female extending about
three times length of head, the 2nd joint obliquely porrect and
fringed with hair above, the 3rd long, naked, and downcurved;
frons smooth ; antennæ simple; tibiæ naked.   Fore wing with
the costa arched at base ; the apex acute and falcate ; the outer
margin excised from apex to vein 5, where it is strongly excurved ;
vein 3 from before angle of cell ; 4, 5 from angle ; 6 from upper
angle; 7, 8, 9 stalked; 10, 11 free; male with a glandular
swelling at base of costa below fringed with long hair and enclosing
masses of flocculent scales.   Hind wing with vein 3 from near
angle of cell ; 4, 5 from angle ; 6, 7 from upper angle ; 8 free.

Fig. 61.

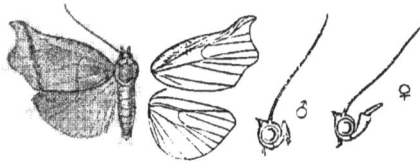

Curicta oppositalis, ♂. ⅓.

Type. (1)†CURICTA OPPOSITALIS, Wlk. xxxiv. 1130.   Salawati; Waigiou;
    Goossensia cinnamomealis, Snell. Tijd. v. Ent.   N. Guinea.
    xxxvii. p. 74, pl. 3. ff. 5, 6.

(2)*CURICTA LUTEALIS, Snell. Tijd. v. Ent. xxxvii. p. 75, pl. 3.
    ff. 7, 8.   Obi.

Genus MURGISCA.

Murgisca, Wlk. xxvii. 11 (1863).

Palpi rostriform, downcurved, thickly scaled, and extending
about three times the length of head; frons smooth; antennæ of
female almost simple; tibiæ smooth.   Fore wing with the costa
nearly straight; the apex acute; vein 3 from near angle of cell;
4, 5 from angle; 6 from near upper angle; 7 given off from 8
before 9; 10, 11 free, or 10 stalked with 7, 8, 9.   Hind wing with
vein 3 from before angle of cell; 4, 5 from angle; 6, 7 stalked,
7 anastomosing with 8.

Fig. 62.

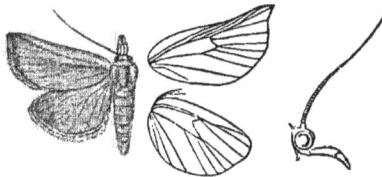

Murgisca cervinalis, ♀. ⅓.

Type. †MURGISCA CERVINALIS, Wlk. xxvii. 12.   St. Domingo.

Genus STREPTOPALPIA.

Streptopalpia, Hmpsn. A. M. N. H. (6) xvi. p. 345.

Palpi rostriform, downcurved, somewhat roughly scaled, and
extending about twice the length of head; frons with a slight
tuft; antennæ of male minutely serrate and ciliated; mid and
hind tibiæ with tufts of hair at middle and extremity, the tarsal
joints tufted.   Fore wing with the apex produced, acute, and
depressed; two tufts of scales on inner margin; vein 3 from near
angle of cell; 4, 5 from angle; 7, 8, 9, 10 stalked, 7 being given
off close to the margin.   Hind wing with vein 3 from near angle
of cell; 4 absent; 6, 7 from upper angle, 7 anastomosing with 8.

684 SIR G. F. HAMPSON ON THE [June 1,

Fig. 63.

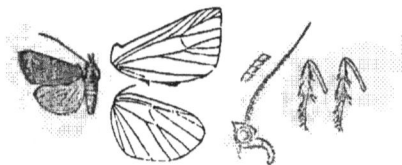

Streptopalpia deera, ♂. ⅔.

Type. STREPTOPALPIA DEERA, Druce, Biol. Centr.-Am., Het. ii. p. 195, pl. 60. f. 1. Mexico; W. Indies.
† „ *ustalis,* Hmpsn. A. M. N. H. (6) xvi. p. 346.

Genus CHALINITIS.

Chalinitis, Rag. Ann. Soc. Ent. Fr. 1890, p. 528.

Palpi rostriform, downcurved, thickly scaled, and extending about twice the length of head; frons with a tuft of scales; antennæ of female minutely ciliated; legs somewhat hairy. Fore wing with the costa straight; the apex rectangular; vein 3 from near angle of cell; 4, 5 from angle; 6 from below upper angle; 7, 8, 9, 10 stalked, 7 being given off before 9. Male with a tympanic vesicle at base of costa of fore wing above, and glandular swelling fringed with long hair below. Hind wing with vein 3 from before angle of cell; 4, 5 from angle; 6, 7 from upper angle, 7 anastomosing with S.

Fig. 64.

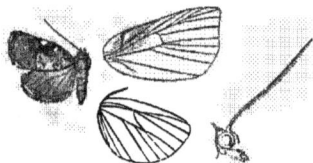

Chalinitis proclea, ♂. ¼.

Type. (1)*CHALINITIS OLEALIS, Rag. Ann. Soc. Ent. Fr. 1890, p. 529, pl. 5. f. 6 (♀). U.S.A.

(2) CHALINITIS PROCLEA, Druce, Biol. Centr.-Am., Het. ii. p. 191, pl. 59. f. 15. Mexico; W. Indies.
†*Torda leucospilalis,* Hmpsn. A. M. N. H. (6) xvi. p. 345.

(3)*CHALINITIS CECROPIA, Druce, Biol. Centr.-Am., Het. ii. p. 191, pl. 59. f. 16. Mexico; Guatemala.

Genus OCRESIA.

Ocresia, Rag. Ann. Soc. Ent. Fr. 1890, p. 485.

Palpi rostriform, downcurved, slightly fringed with hair above,

and extending about four times length of head ; frons smooth ;
antennæ of female almost simple ; legs smoothly scaled. Fore
wing with the costa nearly evenly arched ; the apex produced and
acute ; the outer margin strongly angled at middle ; vein 3 from
near angle of cell ; 4, 5 from angle ; 6 from upper angle ; 7, 8, 9
stalked ; 10, 11 free. Hind wing with vein 3 from before angle
of cell ; 4, 5 from angle ; 6, 7 from upper angle, 7 anastomosing
with 8.

Fig. 65.

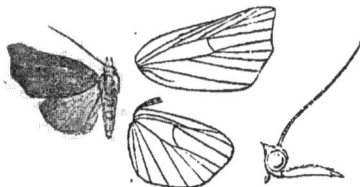

Ocresia bisinualis, ♀. ¼.

Type. *OCRESIA BISINUALIS, Rag. Ann. Soc. Ent. Fr. 1890, p. 486.

Brazil.

Genus PACHYPALPIA.

Pachypalpia, Hmpsn. A. M. N. H. (6) xvi. p. 345.

Palpi rostriform, downcurved, with curled hair at end of 2nd
joint, and extending about the length of head ; frons with a slight
tuft ; antennæ of male thickened and flattened ; mid and hind
tibiæ slightly fringed with hair, the 1st joint of tarsus with a
large tuft of hair. Fore wing with the apex somewhat produced
and acute ; the outer margin slightly angled at vein 4 ; veins
3, 4, 5 from close to angle of cell ; the discocellulars highly
angled ; 7, 8 stalked from 9 ; 10 absent. Hind wing with the
outer margin slightly angled at vein 3 ; vein 3 from near angle of
cell ; 4, 5 from angle ; 6, 7 from upper angle, 7 anastomosing
slightly with 8.

Fig. 66.

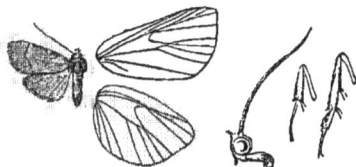

Pachypalpia dispilalis, ♂. ¼.

Type. †PACHYPALPIA DISPILALIS, Hmpsn. A. M. N. H. (6) xvi. p. 345.

W. Indies.

Genus EPITAMYRA.

Epitamyra, Rag. Ann. Soc. Ent. Fr. 1890, p. 503.
Proropera, Warr. A. M. N. H. (6) xvii. p. 453 (1896).

Palpi rostriform, downcurved, smoothly scaled, and extending about twice the length of head; antennæ of female ciliated; tibiæ smooth. Fore wing with the outer margin angled at middle; vein 3 from before angle of cell; 4, 5 from angle; 6 from upper angle; 7, 8, 9 stalked; 11 free. Hind wing with vein 3 from before angle of cell; 6, 7 from upper angle, 7 anastomosing with 8.

Fig. 67.

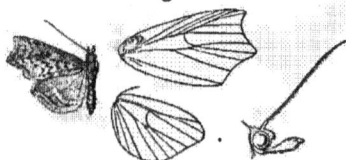

Epitamyra vinosalis, ♂. ¹⁄₁.

SECT. I. Fore wing with vein 10 stalked with 7, 8, 9.

Type. (1)*EPITAMYRA ALBOMACULALIS, Möschl. Lep. Porto Rico, p. 278.
W. Indies.

(2)*EPITAMYRA MINUSCULALIS, Möschl. Lep. Porto Rico, p. 278.
W. Indies.

SECT. II. Fore wing with vein 10 stalked with 11 and anastomosing with 8, 9.

(3)†EPITAMYRA BIRECTALIS, n. sp.

♀. Head and thorax rufous; abdomen ochreous brown. Fore wing rufous, with nearly straight, pale fuscous-edged ante- and post-medial lines; a pale speck on costa before apex; cilia ochreous from apex to the angle at vein 4. Hind wing fuscous, with some red suffusion on outer area at vein 3; cilia from apex to vein 3 red at bases, ochreous at tips.

Hab. Santa Lucia. *Exp.* 20 mm.

SECT. III. (*Proropera*). Fore wing with vein 10 free.

(4) EPITAMYRA VINOSALIS, Warr. A. M. N. H. (6) xvii. p. 454.
Assam.

Genus NACHABA.

Nachaba, Wlk. xix. 834 (1857).
Ascha, Wlk. xxx. 1015 (1864).

Palpi rostriform, downcurved, extending about twice the length of head, and smoothly scaled, the 2nd joint fringed with scales above; frons with a large tuft; antennæ of male with long bristles; mid and hind tibiæ moderately fringed with hair. Fore

wing of male with a tuft of hair on middle of costa above ; underside with a circular flap of scales at base of costa ; vein 2 from
near angle of cell; 3 from angle ; 4, 5 stalked : 6 from below
upper angle ; 7, 8 from angle; 9, 10 absent; 11 free. Hind
wing with vein 3 from angle of cell ; 4, 5 stalked (sometimes
shortly); 6, 7 from upper angle, 7 anastomosing with 8.

Fig. 68.

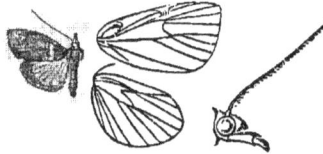

Nachaba flavisparsalis, ♂. ¼.

Type. (1)*NACHABA CONGRUALIS, Wlk. xix. 835.   Brazil.

(2)†NACHABA AURITALIS, Wlk. xix. 834 (*nec* Hübn.).   Brazil.

(3)*NACHABA OPPOSITALIS, Wlk. xix. 835.   Brazil.

(4)†NACHABA FLAVISPARSALIS, Warr. A. M. N. H. (6) vii. p. 424.
                                                            Brazil.
(5)†NACHABA RECONDITANA, Wlk. xxx. 1016.   Brazil.

(6) NACHABA TRYPHŒNALIS, Feld. Reis. Nov. pl. 132. f. 17.   Brazil.
    †   „   *carbonalis*, Warr. A. M. N. H. (6) vii. p. 423.

(7)*NACHABA FUNEREA, Feld. Reis. Nov. pl. 134. f. 23 (♀). Brazil.

Genus SEMNIA.

Semnia, Hübn. Verz. p. 353 (1827).
Acronolepia, Westw. Zool. Journ. v. p. 451 (1834).
Episemnia, Rag. Ann. Soc. Ent. Fr. 1890, p. 481.

Palpi downcurved, slender, extending about twice the length of
head, the 2nd joint fringed with hair above, the 3rd long ; frons with

Fig. 69.

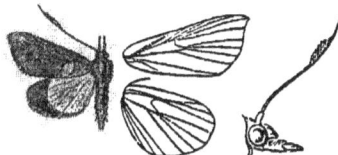

Semnia auritalis, ♂. ¼.

a sharp tuft ; antennæ of male with a thick brush of scales near
extremity ; hind tibiæ slightly fringed with hair. Fore wing with
the costa nearly straight ; the apex rectangular ; male with a
45*

costal fold fringed with hair below; veins 3, 4, 5 from angle of
cell; male with 6, 7, 8, and 10 from upper angle; 9 absent;
female with 7, 8, 9 stalked. Hind wing with veins 3, 4, 5 from
angle of cell; 6, 7 from upper angle, 7 anastomosing with 8.

Type. (1) SEMNIA AURITALIS, Hübn. Zutr. ii. 28, ff. 361, 362.   Brazil.
    Acronolepia quadricolor, Westw. Zool. Journ. 1834, v. p. 451.
        ,,   *biguttalis,* Feld. Reis. Nov. pl. 134. f. 18.
    † *Virbia notata,* Wlk. ii. 472.
    Noctua elongata, Sepp, Surinam, p. 93, pl. 43.
    Episemnia subauritalis, Rag. Ann. Soc. Ent. Fr. 1890, p. 482 ;
        1891, pl. 16. f. 10.

(2)*SEMNIA AURIVITTA, Feld. Reis. Nov. pl. 134. f. 20 (♀).
                                          Brazil.

(3) SEMNIA JOSIALIS, Feld. Reis. Nov. pl. 134. f. 24.   Brazil.

Auctorum.

Episemnia ligatalis, Druce, Biol. Centr.-Am., Het. ii. p. 189, pl. 29.
   f. 11.   Mexico.

Genus EURYPTA.

Eurypta, Led. Wien. ent. Mon. 1863, p. 334.
Chrysophila, Hübn. Zutr. iii. 20 (1825), non descr.

Palpi downcurved, slender, projecting about twice the length of
head, smoothly scaled and fringed with hair below; frons with a
sharp tuft; antennæ of male bipectinated; tibiæ smoothly scaled.
Fore wing with the costa arched at base and towards apex; male
with a slight fold at middle of costa fringed with hair; a glandular
swelling at base of costa below; vein 3 from before angle of cell;
4, 5 from angle; 6, 7, 8, and 10 from upper angle; 9 absent;
in female 7, 8, 9 stalked. Hind wing with veins 3, 4, 5 from
angle of cell; 6, 7 from upper angle, 7 anastomosing with 8.

Fig. 70.

Eurypta basilinealis, ♂ . ¼.

Type. (1) EURYPTA AURISCUTALIS, Hübn. Zutr. ff. 465, 466.   Brazil.
      ,,   *atridorsalis,* Rag. Ann. Soc. Ent. Fr. 1890, p. 481.

(2)†EURYPTA BASILINEALIS, Warr. A. M. N. H. (6) vii. p. 423.
                                          Brazil.

Auctorum.

Eurypta rectibasalis, Rag. Ann. Soc. Ent. Fr. 1891, p. 610.   Brazil.
=   *auriscutalis*, Led. Wien. ent. Mon. 1863, pl. 6. f. 5.
    (*nec* Hübn.).

Genus AROUVA.

Arouva, Wlk. xxx. 963 (1864).

Palpi downcurved, slender, extending once to twice the length of head, and almost smoothly scaled ; frons with a sharp tuft ; mid and hind tibiæ slightly fringed with hair.   Fore wing with the costa straight ; the apex rectangular ; male with a flap of scales on median nervure ; a costal fold below fringed with scales covering a fovea in cell ; veins 3, 4, 5 from angle of cell ; 6, 7, 8, 10 from upper angle, 9 absent ; in female 7, 8, 9 stalked.   Hind wing with veins 3, 4, 5 from angle of cell ; 6, 7 from upper angle, 7 anastomosing with 8.

Fig. 71.

Arouva mirificana, ♂ . ¹⁄₁ .

Type. (1)†AROUVA MIRIFICANA, Wlk. xxx. 963.   Brazil.
    Semnia ægialis, Feld. Reis. Nov. pl. 134. f. 19.

(2)*AROUVA ALBIVITTA, Feld. Reis. Nov. pl. 134. ff. 21, 22.   Brazil.

Genus PENTHESILEA.

Penthesilea, Rag. Ann. Soc. Ent. Fr. 1890, p. 493.

Palpi rostriform, downcurved, thickly scaled, and extending about twice the length of head ; frons smooth ; antennæ of male

Fig. 72.

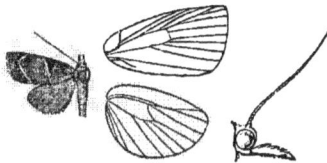

Penthesilea sacculalis, ♂ . ¹⁄₁ .

ciliated ; abdomen with terminal and paired lateral anal tufts. Fore wing with the costa arched at base, then straight, the apex rectangular ; male with a tympanic vesicle at base of costa ; veins

3, 4, 5 well separated at origin; 6 from well below upper angle; 7, 8, 9 stalked; 10, 11 free. Hind wing with veins 3, 4, 5 widely separated at origin; 6, 7 from upper angle, 7 anastomosing with 8.

Type. *PENTHESILEA SACCULALIS, Rag. Ann. Soc. Ent. Fr. 1890, p. 493.
                                                            ? U.S.A.

Genus LOPHOPLEURA.

Lophopleura, Rag. Ann. Soc. Ent. Fr. 1890, p. 506.

Palpi porrect, straight, thickly scaled, and extending slightly beyond the frons, which has a large tuft of hair; antennæ of male ciliated; tibiæ slightly fringed with hair on outer side. Fore wing of male with a glandular swelling at base of costa below fringed with long hair; a fringe of hair on median nervure; the apex rounded; veins 3, 4, 5 stalked; 6 from upper angle; 9 absent; 10 from angle; 11 absent. Hind wing with vein 3 from near angle of cell; 4, 5 from angle; 7 anastomosing strongly with 8.

Fig. 73.

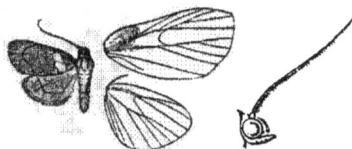

Lophopleura xanthotænialis, ♂. ½.

SECT. I. Fore wing of male with vein 7 stalked with 8; hind wing with veins 6, 7 stalked.

Type. (1) LOPHOPLEURA XANTHOTÆNIALIS, Rag. Ann. Soc. Ent. Fr. 1890, p. 506.   Brazil.
    †*Dastira imitatrix*, Warr. A. M. N. H. (6) vii. p. 425.

SECT. II. Fore wing of male with vein 7 stalked with 6; hind wing with veins 6, 7 from angle of cell.

A. Fore wing of male with no postmedial tuft of scales on costa.

(2)†LOPHOPLEURA SUBLITURALIS, Warr. A. M. N. H. (6) vii. p. 424.
                                                            Brazil.

B. Fore wing of male with a postmedial tuft of scales on costa.

(3)†LOPHOPLEURA EURZONALIS, n. sp.

♂. Dark purplish red-brown. Fore wing with broad antemedial bright yellow band with metallic blue scales on its edges; an indistinct postmedial line angled on vein 6; the margin of both wings suffused with purple and with a series of dark striæ; hind wing fuscous brown, with dark submarginal mark on vein 2.

Hab. Amazons (*Trail*). *Exp.* 18 mm.

Genus CHRYSAUGE.

Chrysauge, Hübn. Samml. exot. Schmett. ii., Lep. iv. Noct. iii. (1806).

Xanthiris, Feld. Wien. ent. Mon. 1863, p. 230.

Candisa, Wlk. xxxiv. 1493 (1865).

Palpi porrect, straight, and hardly reaching beyond the frons, which has a large tuft of hair; antennæ of male almost simple : tibiæ smoothly scaled.   Fore wing with the costa evenly arched ; the apex rounded ; male with a tuft of hair from costa beyond middle, recurved over the wing ; female with veins 7, 8, 9 stalked. Hind wing with vein 3 from near angle of cell ; 4, 5 from angle ; 6, 7 from upper angle, 7 anastomosing with 8.

Fig. 74.

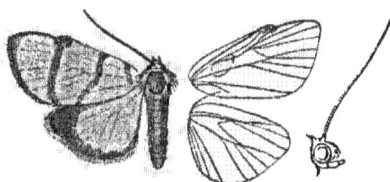

Chrysauge bifasciata, ♂. ⅟.

SECT. I. Fore wing with veins 4, 5 from cell in both sexes ; male with a large fovea covered with hair in cell below ; veins 6, 7 stalked, 8, 9, 10 stalked.

(1)†CHRYSAUGE BIFASCIATA, Wlk. ii. 368 ; Led. Wien. ent. Mon. 1863, pl. 6. f. 1.   Brazil.

(2)†CHRYSAUGE CATENULATA, Warr. A. M. N. H. (6) vii. p. 423.
                                                                   British Guiana ; Brazil.

(3) CHRYSAUGE KADENII, Led. Wien. ent. Mon. 1863, p. 163, pl. 6. f. 2.   Brazil.

(4)†CHRYSAUGE LATIFASCIATA, Warr. A. M. N. H. (6) vii. p. 423.
                                                                   Brazil.

SECT. II. Fore wing with veins 4, 5 stalked in both sexes ; male with no fovea in cell ; veins 8, 9 absent ; 10 free.

Type. (5) CHRYSAUGE FLAVELATA, Cram. Pap. Exot. iv. p. 112, pl. 348. f. B.   Surinam ; Venezuela ; Brazil.
          „   *divida*, Hübn. Samml. exot. Schmett. ii. Lep. iv.
                            Noct. iii.
†   „   *chrysomelas*, Wlk. ii. p. 369.
†*Candisa auriflavalis*, Wlk. xxxiv. 1494.

Auctorum.

Chrysauge unicolor, Berg. Ann. Soc. Arg. xix. p. 274.   Argentina.
Flavinia gopala, Doguin, Le Nat. 1891, p. 109.   Venezuela.

Genera auctorum.

Cryptocosma perlalis, Led. Wien. ent. Mon. 1863, p. 56, pl. 7.
f. 11.   Brazil.
Fenaria sevorsa, Grote, Pap. ii. p. 132.   U.S.A.

Species omitted.

Ethmistis eucarta, Feld. Reis. Nov. pl. 136. f. 28, belongs to the
Pyraustinæ.
Idneodes tretopteralis, Rag. Ann. Soc. Ent. Fr. 1890, p. 605,
probably belongs to the *Schœnobiinæ.*

3. On a Collection of Lepidoptera obtained in the Arusa
Galla Country in 1894 by Mr. F. Gillett. By ARTHUR
G. BUTLER, Ph.D., F.L.S., F.Z.S., &c., Senior Assistant-
Keeper, Zoological Department, British Museum.

[Received May 10, 1897.]

So little has been published respecting the Lepidopterous fauna
of the country south of Shoa, that the present collection, although
unhappily in very poor condition, is of considerable interest [1].
The following is a list of the species :—

RHOPALOCERA.

1. Limnas chrysippus, *L.,* var. klugii.   Between 25th September & 1st October.
2. Ypthima asterope, *Klug.*   „   „
3. Charaxes neanthes ♂, *Hewits.*   „   „
4. Hypolimnas misippus, *L.,* var. inarin. „   „
5. Junonia sesamus, *Trimen.*   Between 1st October & 19th November.
6.   „   octavia, *Cram.*   „   „
7.   „   cloantha, *Cram.*   „   „
8.   „   terea, *Drury.*   Between 25th September & 19th November.
9.   „   cebrene, *Trimen.*   Between 25th September & 21st November.
10. „   clelia, *Cram.*   Between 25th September & 1st October.
11. „   boöpis. *Trimen.*   Between 1st October & 19th November.
12. „   orthosia, *Godt.*   Between 25th September & 19th November.
13. „   taveta, *Rogenh.*   Between 1st October & 19th November.
14. Pyrameis cardui, *L.*   Between 25th September & 1st October.
15. Eurytela dryope, *Fabr.*   „   „
16. Byblia ilithyia, *Drury.*   Between 25th September & 19th November.
17. „   acheloia, *Wallgr.*   „   „
18. Hamanumida dædalus, *Fabr.*   Between 25th September & 1st October.
19. Neptis agatha, *Cram.*   Between 1st October & 19th November.
20. Atella phalantha, *Drury.*   „   „
21. Acræa lycia, *Fabr.,* var. usagaræ.   „   „
22. „   seis, *Feisth.*   Between 25th September & 19th November.
23. Pardopsis punctatissima, *Boisd.*   Between 25th September & 1st October.
24. Polyommatus bœticus, *L.*   Between 1st October & 19th November.
25. Catochrysops asopus, *Hopff.*   21st November.
26. „   osiris, *Hopff.*   „

---

[1] Mr. Gillett says that the collection was made at a place called Sheik
Husein, long. about 40° 45′ E., lat. 7° 44′ S., which accounts for the butterflies
being partly Abyssinian and partly Somalian.

www.ingramcontent.com/pod-product-compliance
Lightning Source LLC
Chambersburg PA
CBHW022012190326
41519CB00010B/1489